Katharina Hertkorn

Shared Grasping:
A Combination of Telepresence and Grasp Planning

Shared Grasping:
A Combination of Telepresence and Grasp Planning

by

Katharina Hertkorn

KIT Scientific Publishing

Dissertation, Karlsruher Institut für Technologie (KIT)
Fakultät für Informatik, 2015

Tag der mündlichen Prüfung: 13. Juli 2015
Referenten: Prof. Dr.-Ing. Tamim Asfour, Prof. Dr.-Ing. Gerd Hirzinger

Impressum

KIT Scientific Publishing

Karlsruher Institut für Technologie (KIT)
KIT Scientific Publishing
Straße am Forum 2
D-76131 Karlsruhe

KIT Scientific Publishing is a registered trademark of Karlsruhe Institute of Technology. Reprint using the book cover is not allowed.

www.ksp.kit.edu

Print on Demand 2016

ISBN 978-3-7315-0402-3
DOI 10.5445/KSP/1000047718

Shared Grasping: a Combination of Telepresence and Grasp Planning

Zur Erlangung des akademischen Grades eines
Doktors der Ingenieurwissenschaften
(Dr.-Ing.)

bei der Fakultät für Informatik
des Karlsruher Instituts für Technologie (KIT)

genehmigte
Dissertation

von

Dipl.-Ing. Katharina Hertkorn
aus Burghausen

Datum der mündlichen Prüfung:	13. Juli 2015
Erster Gutachter:	Prof. Dr.-Ing. Tamim Asfour
Zweiter Gutachter:	Prof. Dr.-Ing. Gerd Hirzinger

Thank you

This thesis has emerged during my employment at the Institute of Robotics and Mechatronics at the German Aerospace Center (DLR). First of all, I would like to thank the former and the current directors of the Institute, Prof. Gerd Hirzinger and Prof. Alin Albu-Schäffer for their support and their great effort to provide a fruitful working environment. I would also like to thank Prof. Tamim Asfour from the Karlsruhe Institute of Technology (KIT) for giving me advice and continuous support in finishing this thesis. I am very grateful to my Department Head, Christoph Borst, for great discussions and the invaluable freedom he gave me by backing me up and believing in me.

I would like to thank all my colleagues at DLR for the delightful atmosphere in the Institute and wonderful summer barbecues and Christmas celebrations. The long list of most-important-people starts with my teammates in the telepresence group and my former team leader Carsten Preusche. In particular, I am deeply grateful to Philipp Kremer, Thomas Hulin, Mikel Sagardia, Bernhard Weber, and Simon Schätzle for everything I learned from them, the wonderful joint work we achieved, the countless hours in the lab, at fairs, demos, and conducting user studies, and all the laughter. I started working in the group during my diploma thesis with a theoretical background in control and knowing only little about practical robotics. Your patience in teaching me everything I needed to know from real-time systems, emergency stops, and calibrating a robot to resolving weird sounds from the joint control of the robot is unforgotten.

This thesis is not only about teleoperation but also about grasping and so I want to thank all the colleagues working in the field of grasping in the Institute. The uncountable discussions with Máximo Roa are the foundations for this work and carried me through all ups and downs during the thesis. Thank you for all the ideas, general advice, indispensable help, and the long late night work on publications. Also many thanks to Theodoros Stouraitis, Bao-Anh Dang-Vu, Thomas Wimböck, Zhaopeng Chen, and Neal Lii.

Numerous is also the help I've received in order to realize the shared grasping approach on the real robot. Particularly, I would like to thank Benedikt Pleintinger, Alex Beyer, Robert Burger, and Florian Schmidt for repairing the hardware when I broke it and developing the low- and mid-level control software. They belong to the backbone of the Institute and without them, all the wonderful robots could not be realized. Also, thanks to Manuel Brucker and Stefan Fuchs who showed the robot how to perceive and understand its environment and I would like to thank Jörn Vogel for the great collaboration in assistive robotics.

Due to all the support that is needed to realize shared autonomy in a complex telepresence system, I will write this thesis in the "we" perspective to honor my colleagues.

In deep gratitude, I also want to thank my parents Johann and Elisabeth, my boyfriend Moritz, my siblings Magdalena and Tobias, and my best friend Inken for their love, empathy, and encouragement. Without them, this thesis would have been impossible and without their good mood, inspirations, support, and patience I would have thrown in the towel a long time ago. Thank you!

Munich, 2015 Katharina Hertkorn

Abstract

Telepresence systems can be used for repair and maintenance in environments that are hard to access or dangerous for the human. Current systems can effectively the gross motor skills of the human operator into actions in the remote environment. In comparison, fine manipulation, like the secure grasping and handling of complex objects, is error-prone and strenuous for the operator. To execute such tasks in a more precise and robust way, this thesis increases the level of autonomy of a complex telepresence system in order to help the operator during the interaction with the remote environment.

The focus of the work is to assist teleoperated grasp and manipulation tasks, which the operator should be able to execute with the lowest possible workload. Therefore, immersion is crucial, and hence, the developed approach does not decrease the high immersion of the human in a telepresence system, but retains important properties like haptic force feedback and 3D stereo view on the remote environment. The operator stays in the loop during prehension and participates actively by commanding continuously the position of the end effector and hence, the hand position from where the object is grasped.

To realize this seamless integration of semi-autonomy in a telerobotic system, first, a new grasp planner is developed. It computes physically feasible and robust grasps fast enough to cope with the movement speed of the operator. The main feature of the approach is, next to the low computation time, the consideration of contact forces that the robotic end effector is able to apply on the surface of the object. Hence, the assumption of normalized contact forces, common in state of the art approaches, is overcome and the grasp can be evaluated realistically with respect to external disturbance forces. Second, an assistance for positioning the end effector relative to the object is developed, which combines a dense database of grasps with the concept of adaptive virtual fixtures. This enables the operator to determine the approach direction towards the object according to the task at hand and additionally, he can continuously influence the selection and execution of grasps.

Besides an objective analysis of both developed assistances, the new telepresence system with shared autonomy is also evaluated with psycho-physical user studies. The human machine interface of the system is enhanced such that the semi autonomous function can be used with haptic force feedback and visual assistance.

The conducted user studies confirm the advantages of the new approach of shared autonomy in a complex telepresence system, which does not reduce the high immersion of the human. The newly developed assistances for grasp and manipulation tasks can be used intuitively by the operator and lead to a considerably more robust and faster execution of the tasks than using direct teleoperation. Additionally, the workload of the operator is reduced when using the new, shared autonomy system.

Kurzfassung

Telepräsenzsysteme können für Reparaturen oder Instandhaltung in Umgebungen eingesetzt werden, die schwer zugänglich sind oder Gefahren für den Menschen bergen. Aktuelle Systeme können die grobmotorischen Fähigkeiten des menschlichen Bedieners gut in der entfernten Umgebung umsetzen. Im Gegensatz dazu ist feinmotorische Manipulation, wie das sichere Greifen und Handhaben komplexer Objekte, fehleranfällig und für den Operator anstrengend. Um solche Aufgaben präziser und robuster zu gestalten, wird in dieser Arbeit der Grad an Autonomie eines komplexen Telepräsenzsystems gesteigert, um dem Operator während der Interaktion mit der entfernten Umgebung zu assistieren.

Der Fokus der Arbeit liegt auf der Unterstützung von teleoperierten Greif- und Manipulationsaufgaben, die mit möglichst geringer Operatorbelastung durchgeführt werden sollen. Dazu ist Immersion erforderlich, weswegen das entwickelte Konzept die hohe Immersion des Menschen in Telepräsenzsystemen nicht verringert, sondern wichtige Eigenschaften, wie z.B. Kraftrückkopplung und 3D Sicht auf die entfernte Umgebung, erhält. Der Operator bleibt im Greifvorgang eingebunden und kann ihn kontinuierlich und intuitiv steuern und beeinflussen, da er weiterhin die Position des Endeffektors bestimmt und damit die Handposition, aus der das Objekt gegriffen wird.

Um diese nahtlose Integration von Teilautonomie in ein Telerobotiksystem zu verwirklichen, wird als erstes eine neue Methode zur Greifplanung entwickelt, die physikalisch realisierbare und robuste Griffe schnell genug berechnet, um mit den Bewegungen des Operators mithalten zu können. Die Besonderheit des Ansatzes ist neben der geringen Berechnungszeit die Berücksichtigung von Kontaktkräften, die der robotische Endeffektor tatsächlich aufbringen kann. Damit wird die Annahme von normalisierten Kontaktkräften, wie bei bisherigen Ansätzen in der Literatur, überwunden und der Griff kann in Bezug auf externe Störkräfte erstmals realistisch bewertet werden. Als zweites wird eine

Assistenz zur Positionierung des Endeffektors relativ zum Objekt erarbeitet, die das Konzept einer großen Datenbank an Griffen mit adaptiven virtuellen Orientierungshilfen verbindet. Auf diese Weise kann der Operator die Richtung aus der das Objekt gegriffen werden soll je nach Aufgabe selbst bestimmen und die Auswahl und Ausführung der Griffe beeinflussen.

Neben der objektiven Evaluierung der beiden entwickelten Assistenzen, wird das neue teilautonome System zusätzlich anhand psycho-physischer Studien bewertet. Dazu wird die Mensch-Maschine Schnittstelle des Systems so erweitert, dass die teilautonomen Funktionen mit haptischer Rückkopplung und visueller Assistenz verwendet werden können.

Die durchgeführten Benutzerstudien bestätigen den neuen Ansatz von Teilautonomie in einem komplexen Telepräsenzsystem, durch den die hohe Immersion des Menschen nicht verringert wird. Die neu entwickelten Assistenzen für Greif- und Manipulationsaufgaben können vom Operator intuitiv bedient werden und bewirken eine deutlich robustere und schnellere Ausführung der Aufgaben. Zusätzlich wird die Arbeitsbelastung des Operators während der Steuerung des neuen teilautonomen Systems reduziert.

Contents

1 Introduction

During the industrial revolution, the combination of electrics and mechanics allowed the realization of the first automation machines. Ever since, the development of automatisms and robots advanced towards the ultimate goal of a machine that equals, or even outperforms, the human being in terms of physical and cognitive abilities. Although machines might surpass the human in particular skills, this goal is most likely not reachable within one machine for a general variety of tasks. Yet, robots are nowadays part of our every day life. They are used, for example, as fully integrated automation machines in industrial settings, inside our homes as vacuum cleaners, and enrich our professional life as advanced video conferencing tools. Nevertheless, applications that involve physical interaction with the environment are far from being solved by a fully autonomous robot. This especially holds true for general-purpose robots with a high number of degrees of freedom, like a humanoid equipped with robotic hands. Although the robotic hardware itself may be physically capable of performing certain tasks, the complexity of planning and controlling, the demand of cognitive abilities like scene understanding and interpretation, as well as the absence of appropriate reactions in unforeseen circumstances still pose unsolved challenges for the autonomous behavior of these robots.

A widely used technique to overcome these limitations is direct *teleoperation*. It enables a human operator to remotely control a robot via direct coupling and to use the robot as an advanced technological tool to interact with a distant or potentially inaccessible environment. Hence, the robot needs only a low level of autonomy and strictly follows the commands of its operator who makes the cognitive decisions. In telerobotics, where the robot is located a significant distance from the human operator, these decisions are more likely to be appropriate and a high task success rate can be achieved if the operator experiences *telepresence*. This is the feeling of being physically present in the remote environment by means of high immersion, for instance by using multi-modal feedback like simultaneous visual and haptic feedback. Still, it remains a tedious task for the human to manually operate a robotic system from a distance to solve physical interaction tasks like grasping and manipulation, especially if a multi-fingered robotic hand is used. Moreover, training is needed to understand the technical capacities of the hardware and its efficient use. The difficulty of telemanipulation increases especially when appropriate feedback is missing, for example a lack of haptic feedback on the fingers for grasping tasks.

While grasping and manipulation are challenges in telemanipulation, they can also not be solved in a completely autonomous way by the robot due to, for example, missing or incomplete task knowledge, absent reaction to failures and unforeseen events, or lack of understanding of the physical behavior of environmental objects. To bridge this gap between fully autonomous robots on the one side and directly teleoperated robots on the other side, varying levels of autonomy can be realized. These systems are called *shared autonomy* systems, which combine advantages of both direct coupling and full autonomy. In unexpected situations or delicate tasks, the human knowledge can be accessed while repetitive and tedious tasks are automated. Similar to teleoperated systems, the performance of shared autonomy systems also depends on the capabilities of the human operator, and it is therefore important to keep the operator immersed. Then, he stays in-the-loop, is focused, and concentrated. This is, for example, influenced by the distribution of load for the task execution between the autonomous behavior and the human operator, and also the user interface for accessing the autonomous functions.

In this thesis, a shared autonomy system is designed and realized to assist a human operator in telemanipulation tasks in order to make them more precise and robust when compared to direct teleoperation. It is based on a telepresence system that allows the human operator to control a dexterous robot while being highly immersed in the remote environment by multi-modal feedback (Fig-

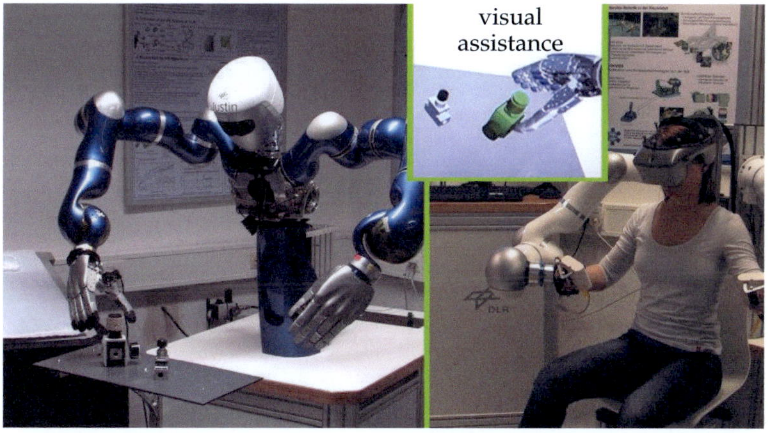

Figure 1.1.: Semi-autonomous grasping using DLR's telepresence system. It consists of the remote robot SpaceJustin (on the left, [27]) and the human-machine interface HUG (on the right, [249]). Visual assistance, additional to the camera stream of the remote environment, is displayed using the head-mounted display.

ure 1.1). In contrast to current shared autonomy systems, the developed system sustains the immersion with important factors like haptic feedback for the hands and hence, keeps the human concentrated and involved in the task. Especially grasping and manipulation of objects with teleoperation systems often results in steep learning curves, grasp failures, and a high workload of the operator due to the shortcomings of teleoperating a multi-fingered robotic hand. Thus, the focus of the thesis is the assistance of grasping and manipulation tasks to overcome drawbacks in the telepresence system such as missing haptic feedback for the fingers.

To realize an immersive shared autonomy system for grasping and manipulation, new autonomous grasp planning methods in combination with a grasp force optimization are developed. Additionally, the interface allows for an intuitive use by the operator by realizing kinesthetic feedback for the arm and an optional force feedback for the fingers with a one degree of freedom device that reflects a force proportional to the grasp forces in the system. Therefore, we developed a system that involves the operator in the grasping process. The overall structure of the presented assistance is derived from human prehension. There, the movement of the human hand can be divided into three steps: hand positioning relative to the object, finger adjustment to the object, and the actual grasp. Each of these steps is assisted with newly developed autonomous functions that can be used in telepresence systems (left and middle part

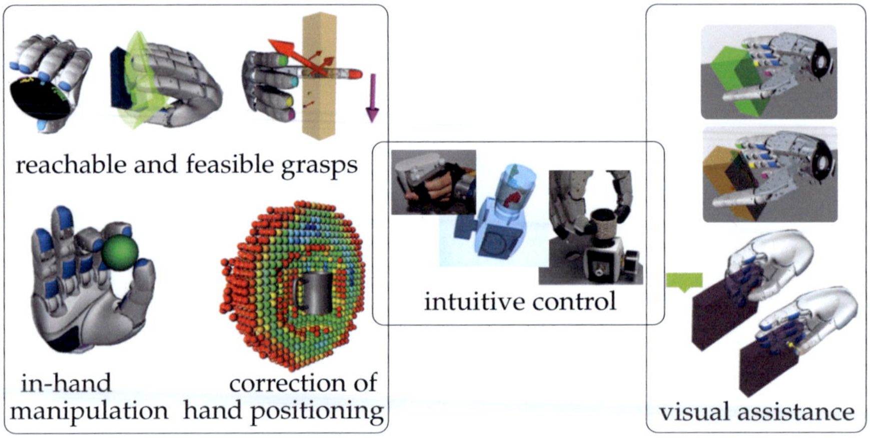

Figure 1.2.: Assistance realized with shared autonomy. Left: semi-autonomous functions, middle: intuitive control of the actual grasp, right: visual assistance for the operator.

of Figure 1.2). As can be seen in Figure 1.1, the operator continuously controls the pose of the end effector while having a stereo view of the remote environment. This setup allows the human to determine and adjust the hand pose to grasp the object. The resulting hand pose is then used by the semi-autonomous functions as basis of computation ensuring a close cooperation between autonomous functions and the human operator. To allow an intuitive use of the autonomous functions by the operator to perform the actual grasp, an interface with haptic feedback and visual assistance is developed (right side of Figure 1.2). All assistances are evaluated with suitable user studies. The remainder of the thesis is structured according to the developed assistance functions and their evaluation, as summarized in Figure 1.3.

Chapter 2: Assistance for Grasping in Telepresence Systems

This chapter offers an initial introduction to teleoperation and telepresence systems. Then, several classifications are introduced to describe the various levels of autonomy, that can be realized in robotic systems. Based on these, the state of the art systems for shared autonomy, used for grasping and manipulation are presented. An overview of human prehension is also provided, as it is the base of the newly developed framework. Each assistance functionality is then presented in its own chapter.

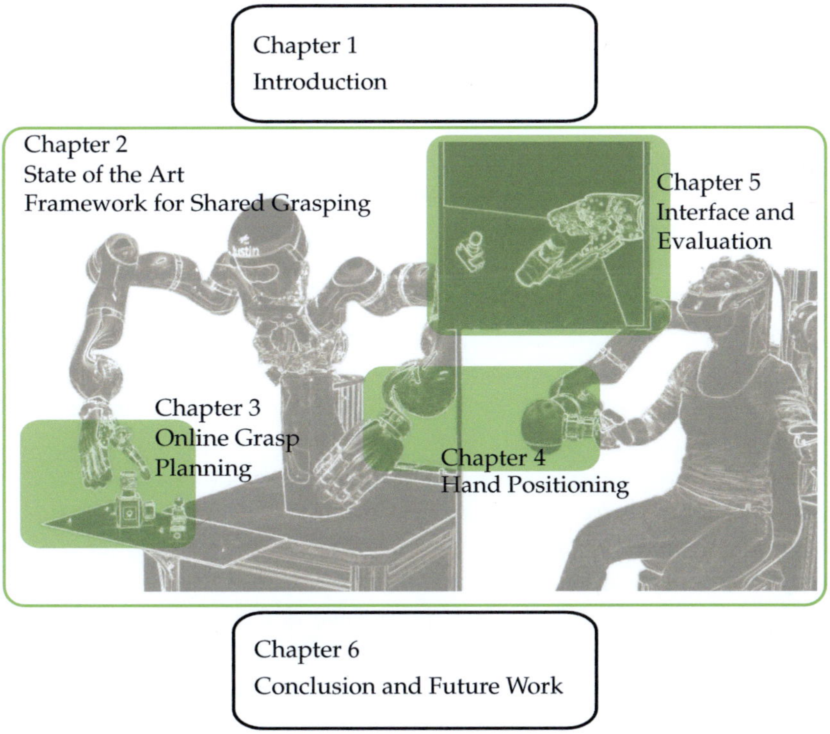

Figure 1.3.: Structure of the thesis.

Chapter 3: Online Grasp Planning

The "finger adjustement" in human prehension corresponds to the autonomous computation of contact points and forces for a robotic hand. It is realized with a newly developed grasp planner that is based on the concept of "Independent Contact Regions" (ICRs). This concept enables finding contact patches on the object rather than single contact points. It improves the robustness of the grasp against positioning errors in the execution with a real system, as the grasp is stable as long as the chosen contact point lies within the contact regions. The concept of ICRs was enhanced by taking into account the joint and torque limits of the robotic hand while still guaranteeing a low computation time. This was achieved by splitting the algorithm into online and offline elements. Offline, the workspaces of the robotic fingers are computed, and they are later used online to obtain reachable regions on the object. Based on these reachable regions, contact regions for a stable grasp are determined. As a grasp not only requires the location of contacts but also suitable finger forces, this thesis presents a method based on "ray-shooting" to simultaneously plan both of them while taking into account the torque limits of the robotic hand. This serves to overcome the assumption of normalized contact forces used in state of the art approaches, and to realistically evaluate the grasp with respect to external disturbance forces. Finally, the method chooses a grasp where the fingers can apply the maximal contact force in their current configuration which is particularly important when heavy objects have to be lifted.

Chapter 4: Guiding the Hand Pose

As a next step, two assistances for positioning the end effector relative to the object are presented. These assistances are particularly important if the kinematic of the end effector is not intuitively understood by the operator, or if the human or the human-machine interface have limited input capabilities. Then, this assistance can be enabled to find a better grasping configuration either with visual assistance or using virtual fixtures. The visual assistance is based on the grasp planning approach presented in Chapter 3, and the operator is led to a valid hand pose relative to the object with a virtual arrow. The assistance with virtual fixtures guides the hand pose toward a valid grasp. This method uses an offline computation of a database of grasps, the so-called "graspability maps". To obtain them, the space around an object is sampled and the valid grasps for a certain approach direction are stored for a hand-object combination. Online, this database is searched for a suitable hand pose relative to the object taking into account the current approach velocity, the current hand pose, and the number of valid grasps for the target pose. The main feature of the developed method is the continuous re-evaluation of the target pose, which allows the operator to withdraw from the object, move around it, and in this way choose the optimal grasp with respect to the task at hand.

Chapter 5: Evaluation of Assistance

In this chapter, the newly developed assistance functionalities are evaluated with respect to the human operator by well-established psycho-physical methods. The assistances are implemented for DLR's telepresence system, and additionally tested in a setup designed for assistive robotics. In both systems, the anthropomorphic DLR-HIT Hand II is used as end effector. It has five fingers with three degrees of freedom each [124]. Three user studies are presented to evaluate the assistances with standard psycho-physical methods, which provide information to select an intuitively understandable visual assistance. The visual assistance continuously displays the results of the grasp planning and additional information, like detected object pose, during prehension. The studies show in a statistically significant way that the assistances reduce the workload of the operator while simultaneously improving the performance of grasping tasks. The visual assistance increases the immersion of the operator into the remote environment.

Chapter 6: Conclusion and Future Works
Finally, the thesis is summarized and future steps are pointed out.

In summary, four major contributions are presented in this thesis:

- A new approach to shared autonomy is developed to be used in a telepresence system for grasping and manipulation tasks in order to reduce the workload of the operator while maintaining the high immersion of the operator into the remote environment. Assistance functions are autonomous grasp planning and guidance for positioning the end effector relative to the object.
- The development of a grasp planning algorithm with low computation time to increase the stability of grasps. The planner takes into account the kinematics of the robotic hand and its torque limits. In contrast to other state of the art approaches, this planner allows to realistically evaluate grasps as it includes the information for the contact forces that the robotic hand can apply on the object with its current configuration.
- The guidance for the position of the end effector is a new method that merges the ideas of adaptive virtual fixtures and a pre-computed grasp database. Due to dynamic recalculation of the target pose, the operator is not limited to a few preselected grasps but can freely move around the object and therefore, explore all possibilities of the database.
- An intuitive interface for the operator is presented, which merges haptic feedback and visual assistance. It is developed based on user studies, which show a reduction in operator workload and an increase in task performance of the system.

This thesis presents only parts of the papers published by the author, and therefore, a complete list of publications ([236–259]) is given at the end of the thesis.

2

Assistance for Grasping in Telepresence Systems

At first, this chapter gives a general introduction to telerobotics and telepresence systems and provides details on the performance measures used to evaluate such human-centered systems in Section 2.1. Challenges in grasping and manipulation tasks are pointed out, especially for systems using direct teleoperation. To overcome drawbacks and limitations in these systems and to support the human operator, they can be enhanced with an autonomous capacity of reaction to the environment. These types of systems are then called *shared autonomy systems*, and they are classified in general in Section 2.2. The state of the art of these systems is described focusing on telerobotic systems for grasping and manipulation. Based on this knowledge, the background for the specific approach developed in this thesis is then given in Section 2.3. We present a method for a shared autonomy system with assisted grasping and manipulation using a multi-finger hand while providing a high level of immersion of the human operator. Thus, we analyze first a control scheme of human prehension in order to identify requirements for an intuitive assistance. As a second step, we present the conceptual approach taken in this thesis for assisted grasping and manipulation in dexterous telerobotic systems.

2.1. The Basics of Telerobotics and Telepresence

In telerobotic systems, the operator is enabled to control a remote robot, the *slave*, by using a human-machine interface, the HMI or *master*, as outlined in Figure 2.1. The commands from master to slave and the feedback from slave to master are transmitted via a communication channel [149]. Telerobotics allows to explore or interact with a remote environment and is used in many applications. The first application of bilateral telerobotics, in the 1940s, was the handling of radioactive material and a wide range of telerobotic systems has been since then developed [14]. They are nowadays available for fields of application like minimal invasive surgery, combat and security, nuclear plants, maintenance and repair, or on-orbit servicing [15, 48, 120, 157, 204]. Shortly, telemanipulation systems are applied in all areas where the environment is potentially dangerous or hard to access for the human.

Although the structure of the systems is always the same, they differ in their embodiment significantly. Figure 2.2 and 2.3 show some examples of currently available master and slave devices. The kinematic capabilities of the HMIs (Figure 2.2) range from a few degrees of freedom, e.g. a computer mouse [218], to highly dexterous input devices that can even display haptic feedback on fingers of the human hand, e.g. Sigma from Force Dimension [50] or CyberForce from CyberGrasp [205]. The remote robots (Figure 2.3) are usually adapted to their application, e.g. a remotely operated vehicle for survey operations in deep sea by Schilling Robotics [176] or the DaVinci Surgical System for minimal invasive surgery [204]. Hence, telerobotic systems exist with a wide range of kinematic

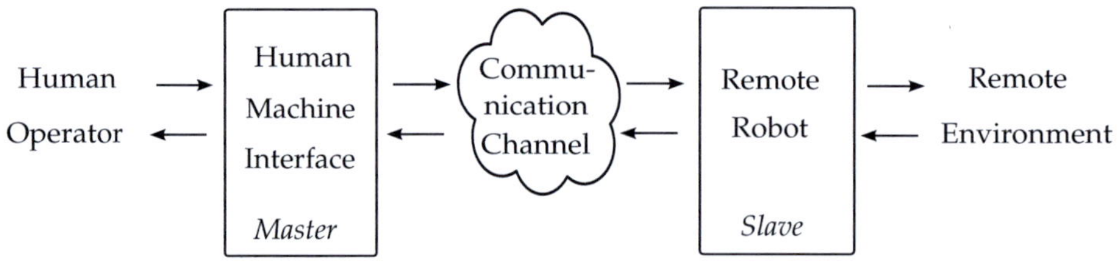

Figure 2.1.: Schematic structure of a telerobotic system.

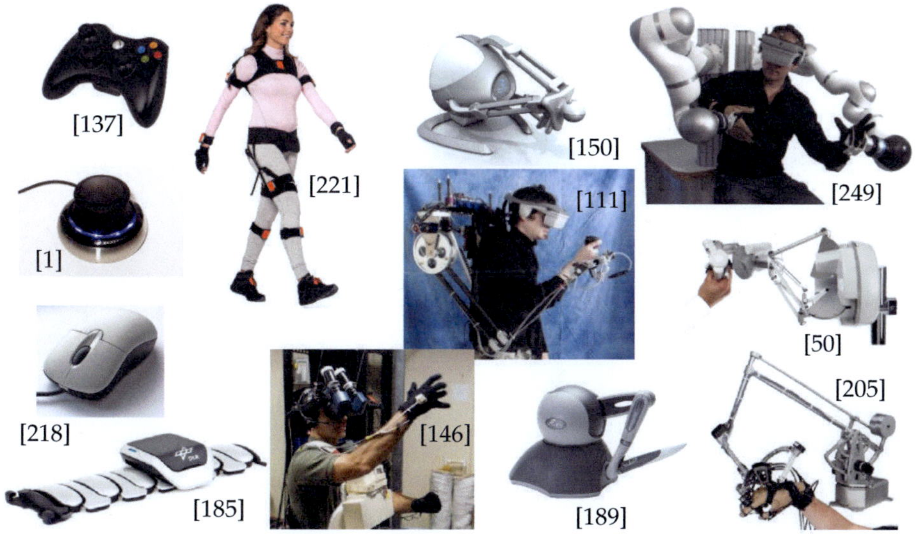

Figure 2.2.: Some examples for human-machine interfaces: a computer mouse [218], a SpaceMouse [1], the XBox Controller [137], a body suit to track the movements of the human body [221], an interface to interact with a virtual environment consisting of several trackers, datagloves, and a head-mounted display [146], a tactile feedback device [185], several haptic feedback devices with varying numbers of degrees of freedom, i.e. the Falcon [150], Phantom Omni [189], Sigma [50], the Cyber Force [205], a wearable haptic device [111], and a bimanual haptic device with a large workspace [249]. The HMIs range from devices with a few degrees of freedom, e.g. [218], to highly dexterous haptic devices, e.g. [249].

and dynamic capabilities including size of workspace, strength, and speed. The conditions in operation lead to telerobotic systems with communication channels ranging from high-bandwidth and almost no time delay, e.g. in minimal invasive surgery, to low-bandwidth and high time delays of several seconds, e.g. in deep sea or space applications.

Major goals of development are to ensure a safe and intuitive handling of the system for the human while simultaneously reaching a high task success rate. Achieving these goals depends on all subsystems of the telerobotic system: the human, the human-machine interface, the properties of the communication channel, the remote robot, and the environment to interact with (Figure 2.1). Any limitation in one of these elements potentially drops performance of the system, as they are tightly coupled, and restricts the tasks that the system can fulfill. In the following, we focus on components that have a large influence on

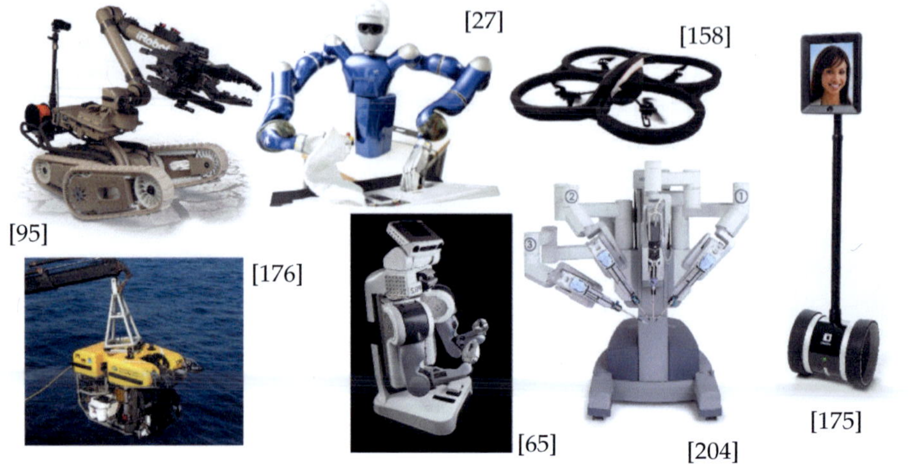

Figure 2.3.: Some examples for remote robots in the fields of minimal invasive surgery [204], inspection [158], deep sea [176], or military intervention [95], general-purposed robots [27, 65], and telepresence robots used as advanced video conferencing tools [175].

the performance of systems designed for manipulation: the human, the dexterity of remote robot and HMI, the capability of the system to display multi--modal feedback enabled by sensor modalities and control, and the mapping between remote robot and HMI.

As the human is a permanent component in telerobotics, his subjective impression of the handling of the system plays an important role and also influences the task success rate. The subjective sense of the operator to feel present in the remote environment is called telepresence and enables intuitive interaction with the remote environment [201]. This mental state is influenced by the characteristics of the system that the human is working with, the task he is performing, and his innate abilities and personal experiences [53]. The term telepresence has been recently employed to also describe enhanced video conferencing tools like the ones commercialized by e.g. iRobot or VGo Communications, where cameras and a screen are attached to a mobile platform [47, 175]. In contrast, this thesis focuses on systems for remote hand-arm manipulation. The ideal state for a technical mediated presence is called transparent when the operator cannot distinguish between the interaction with the distant environment and the interaction displayed by the HMI [201]. Transparency involves the evaluation of different factors that influence people's feeling of presence such as multi-modal presentation and its consistency, immersion and realism [219]. Increasing the sense of presence is therefore assumed to benefit the operator with a decreased workload and an increased focus on the task.

The technological realization of the system influences the subjective impression of the human operator, i.e. by the capabilities of displaying multi-modal feedback, as well as the range of manipulation tasks that the system can perform. In order to create a telerobotic system that is capable of a wide range of manipulation tasks, the remote robot and the HMI need to be dexterous. As pointed out by Shimoga in 1993 [193], a remote robot equipped with a multi-finger hand as end effector is crucial to increase task performance. In comparison to two-finger grippers, a hand allows grasping an object with uneven surfaces more easily, and can perform its reorientations with in-hand manipulation. A hand can even make use of the structural properties of the object, like activating a power button while holding the tool in the hand.

A dexterous robot usually also extends the possibility to sense the remote environment which can lead to multi-modal feedback to the human, and therefore, an increased feeling of telepresence [120]. However, the master device requires therefore at least the same dexterity as the slave. It needs to be able to measure the human motion and display the perceptual information from the remote environment. This perceptual information is usually reduced to visual and haptic feedback in telemanipulation systems, human's most needed modalities to perform effective and useful manipulation [212]. Stereo visual feedback provides the human operator with depth information about the remote environment. It has been found superior to other display types [42, 194], but requires a high-bandwidth communication channel. Additionally, the human needs to use a device that enables impression of stereo like head-mounted displays, autostereoscopic devices, or polarized glasses. Haptic feedback should be displayed on the whole hand of the human, including the fingers, in the context of controlling a dexterous robotic end effector in direct teleoperation in order to increase consistency and realism of the feedback.

An entire research field of *bilateral teleoperation* is dedicated to develop control methods enabling haptic feedback, thus closing the haptic control loop such that a stable and transparent interaction with the remote environment is possible [91]. In this field, major challenges in the communication channel like time delay, jitter, and packet loss are tackled as they can additionally degrade the performance of telemanipulation systems [149]. This research field mainly focuses on enabling haptic feedback for the hand, commonly on one point at the palm. It has still been found to increase system performance significantly compared to systems without haptic force feedback by reducing task completion time, lowering the magnitudes of contact forces, and reducing the errors during the task [42, 77, 216]. Additional haptic force feedback for the fingers is equally important [193] but harder to realize due to the kinematic complexity of the human hand. Without force feedback, participants were found to exert a higher grip force especially when accuracy demands are high [67]. Observing the finger motion of a human hand to command the robotic hand using a vision system is challenging due to occlusions as the human hand can be attached to a haptic device. Hence, sensorized gloves are usually used [51]. Their results highly depend on the calibration routine, which can be quite complex to account e.g. for varying sizes of human hands [62, 74]. Several hand exoskeletons have been developed providing haptic feedback for the fingers but usually without combination of feedback for the hand [84].

Intuitive control of the remote robot depends also on the chosen mapping from human to robot actions. The mapping usually translates human to robotic motion and also influences the force feedback. The simplest form is joint-to-joint mapping, where each joint of master and slave are coupled. This is often used for teleoperating anthropomorphic robotic hands [119], but it does not take the kinematic constraints of the robot into account potentially resulting in unexpected behavior. The most common mapping for the coupling of the arms is point-to-point or Cartesian mapping. A point on the end effector of the master is coupled to a point on the end effector of the slave. For hands, this strategy can be realized by coupling points on the fingertips [159, 177]. In [163], a data driven mapping strategy is developed for the arms which could also be used for hands. There, the human follows the motion of a robot freely, his movements are recorded, and the chosen mapping is applied and then used to control the arm. Coupling predefined poses or gestures has the advantage of not relying on the precise shape of the human hand [45, 122]. Synergies or virtual objects can be used to provide a mapping between human and robotic hand, which results in commands that are independent from the kinematics of the robotic hand [69, 183].

Current telemanipulation systems only partially realize a dexterous, transparent system due to the mechanical complexity of simultaneously displaying haptic feedback for both the hands and the fingers and additionally controlling a multi-finger robotic hand. Instead, finger feedback is either left aside [111], or simplified with the use of a robotic gripper with one degree of freedom (DoF) which can be easily displayed to the human with force feedback at the expense of reduced manipulation capabilities [37, 107]. Robotic hands are usually controlled with a dataglove without haptic feedback for the fingers [19, 128] or without haptic feedback for the wrist [84]. Both options degrade the performance of grasping and manipulation tasks. So, even when dexterous telepresence systems are used, a major loss of dexterity is likely to happen [66].

Performance Measures

As the performance of a telerobotic system depends on all components of the system and their interaction, several evaluation methods have been developed. They usually focus on single characteristics of the overall system. For example, for the control aspects of the direct coupling in teleoperation, transparency is typically used as goal for the developed control. It is analyzed with several analytical measures that result in comparisons between the remote robot and the HMI, like comparing the position and force error for the quality of haptic telepresence [85]. Also, the mapping between master and slave can be evaluated as subpart. Examining for example the mapping between hands, the similarity of certain hand shapes or fingertip positions is measured, dependent on the emphasis of the system [45, 159]. Focusing on the human in the telerobotic system, the subjectively achieved level of telepresence can be evaluated [8]. It can be determined by psycho-physical user studies where participants fill out questionnaires according to their individual feelings [53, 219]. There, multi-modal presentation, its consistency, immersion, and realism are considered as factors that influence the transparency of the system and hence, the feeling of presence. Finally, the task performance can be taken into account by measuring the task completion time, the reaction time to a secondary task, or the task success for certain exercises.

In general, all characteristics of a telerobotic system need to be considered to evaluate its performance. First, the subparts should be analyzed with their specific criteria and then the task performance as well as the subjective sense of telepresence of the operator needs to be taken into account. Therefore, an evaluation of telerobotic systems always needs to incorporate psychophysical user studies with both objective and subjective measures, since the feeling of intuitive interaction depends on the subjective perception of the operator. In general, intuitive interaction can be coupled to the subjective sense of telepresence of the operator [53]. In detail, multi-modal presentation, its consistency, immersion, and realism are considered as factors that influence the transparency of the system and hence, the feeling of presence [219]. Increasing the sense of presence is then assumed to benefit the operator with a decreased workload and an increased focus on the task.

2.2. Classification and Overview of Shared Autonomy

To overcome the drawbacks of direct coupling in teleoperation and telepresence systems, varying levels of autonomy can be introduced. An autonomous robot was defined by Beer et al. [16], which is based on the sense-plan-act paradigm reviewed in [142], as:
*"The extent to which a robot can **sense** its environment, **plan** based on that environment, and **act** upon that environment with the intent of reaching some **task-specific goal** (either given to or created by the robot) without external **control**"* [16].
The level of autonomy of a robotic system towards the completely autonomous robot can be classified from a human- or a robot-centered perspective. The robot-centered approach does not take into account the interaction or supervision of a human. Instead, the robot's autonomy level increases with its abilities to execute tasks, react to failures, and take cognitive decisions. In the human-centered approach, the human is recognized as an important part of the system who can provide the robot with required input to execute tasks. The level of autonomy is then categorized according to the distribution of decisions taken by the human or the robot, or for example according to the amount of time the human or the robot need to execute a task. As we focus on increasing the performance of telepresence system in this thesis, taking the human-centered approach is natural. In the following, a short review on human-centered taxonomies for telerobotic systems is given and the wording is clarified. Then, several state of the art shared autonomy systems, i.e. telerobotic systems that are suitable for manipulation and incorporate some degree of autonomy, are presented and classified according to the chosen taxonomy. After a general overview, we describe systems in particular that use multi-finger hands as end effectors.

2.2.1. Levels of Autonomy

Several human-centered taxonomies describing the levels of autonomy of technical systems have been developed. The earlier classifications are in the context

Table 2.1.: Scale of degrees of automation [191].

1.	The computer offers no assistance, human must do it all.
2.	The computer offers a complete set of action alternatives, and
3.	narrows the selection down to a few, or
4.	suggests one, and
5.	executes that suggestion if the human approves, or
6.	allows the human a restricted time to veto before automatic execution, or
7.	executes automatically, then necessarily informs the human, or
8.	informs him after execution only if he asks, or
9.	informs him after execution if it, the computer, decides to.
10.	The computer decides everything and acts autonomously, ignoring the human.

of automation of industrial processes [191] and focus on the distribution of decisions between human and computer and the amount of information they exchange. At that time, the term *computer* was used instead of *machine* and we keep this terminology to emphasize that the developed classifications focus on the decision making process and explicitely do not take into account a robot. The most descriptive taxonomy was presented by Sheridan in 1992 [191]. It consists of 10 levels of automation (LOAs), summarized in Table 2.1. They range from Level 1, where the human decides everything without assistance of a computer, to Level 10, where the computer is completely in charge and ignores the human. This taxonomy, however, is rather specific for the decision making process and its output. Endsley and Kaber [57] published a more general description of LOAs taking into account four major stages of autonomy: monitoring, generating, selecting, and implementing. Thereby, they broaden the one-dimensional scale of Sheridan to two dimensions, as presented in Table 2.2. In *monitoring*, displays are utilized to perceive the system status which is then used to *generate* options or strategies for achieving goals. Among the options, a particular strategy is *decided* and *implemented*, i.e. executed. This allows to distinguish the roles of human and machine more specifically. Again, the human is in charge of all functions in the lowest level of automation (*manual control*), while for the highest (*full automation*) the computer takes over. In between, the computer gets gradually more sophisticated. For example, Level 4 is named *shared control* where both the human and the computer generate possible decision options. The human still holds full control over the selection of which option to implement; however, carrying out the actions is shared between the human and the computer [57]. Level 9, one level before full automation, is *supervisory control* where the computer generates options, selects them,

Table 2.2.: Two-dimensional hierarchy of levels of automation (LOAs) [57]; H: human, C: computer.

	Level of automation	Monitoring	Generating	Selecting	Implementing
1.	Manual control	H	H	H	H
2.	Action support	H/C	H	H	H/C
3.	Batch processing	H/C	H	H	C
4.	Shared control	H/C	H/C	H	H/C
5.	Decision support	H/C	H/C	H	C
6.	Blended decision making	H/C	H/C	H/C	C
7.	Rigid system	H/C	C	H	C
8.	Automated decision making	H/C	H/C	C	C
9.	Supervisory control	H/C	C	C	C
10.	Full automation	C	C	C	C

and executes them as well. The human only monitors the process and intervenes if necessary. During intervention, the LOA of the system changes, for example to Level 5 where the human can select his own option of action.

Four stages or functions of automation were also identified by Parasuraman et al. [156] but referred to as information acquisition, information analysis, decision and action selection, and action implementation. They proposed to distinguish the LOA in each of the four stages in more detail by using not only H, C, or H/C as levels (like Endsley et al. [57]), but using the 10 LOAs by Sheridan (Figure 2.4). This allows to reflect the various degrees of automation that can be realized in each function. The stages of automation by Parasuraman et al. can be well connected to the highlighted words in the definition of autonomous robots: *sense* (information acquistion), *plan* and *goal* (information analysis and decision selection), and *act* or *control* (action implementation). This indicates that a classification of the autonomy of a robot with a two-dimensional scale is appropriate [16].

In addition to the classification of technical systems in the context of automation, various methods and taxonomies have been developed to categorize levels of autonomy in telerobotic systems [7, 16, 52, 100, 138]. In comparison to the previous taxonomies, the task is always executed by the remote robot as the human has no possibility to interact with the remote environment. However, the systems vary significantly in the capabilities of the robot to react autonomously to its environment or how much it can alter the commands of the human. The level of autonomy of the robot is also correlated to the amount and frequency

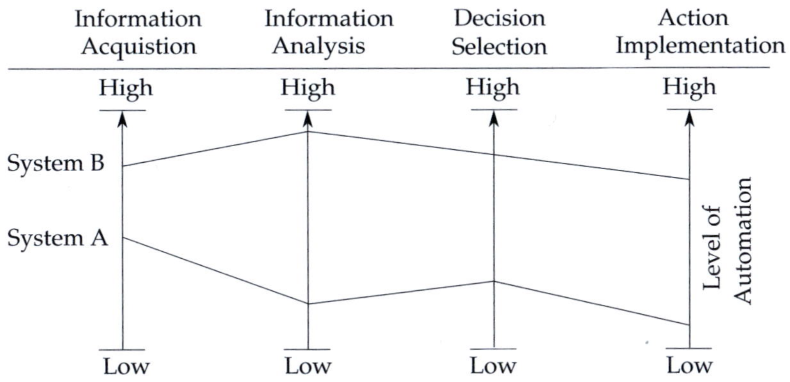

Figure 2.4.: Levels of automation for independent functions of information acquisition, information analysis, decision selection, and action implementation (two examples for potential systems are given) [156].

of information per task needed from the human operator to remotely control the robot in order to fulfill a task. This mirrors the level at which the operator gives input to the remote robot. Unfortunately, the names for the levels of autonomy of telerobotic systems and for the methods to achieve a certain level have not been used consistently in literature. Therefore, we give an overview of the wording in the field in the following and clarify their meaning.

Direct Teleoperation, Direct Control, Remote Programming

Direct teleoperation or *direct control* does not include autonomy of the remote robot, but corresponds to a basic teleoperation system. It describes direct coupling between a human operator and the robot, where the robot strictly follows the continuously given commands by the operator and optionally sends continuously measured sensor signals as feedback. In this concept, the mapping between human and robot can vary, introducing for example a scale in the commanded movement or velocity. This is often used in medical applications to enable very small precise motions of the remote robot [108], or for controlling robotic hands with datagloves choosing an appropriate mapping between human and robotic finger motions [51]. Additionally, impedance control, position interpolation, or adaptations due to joint limits or internal posture optimization (e.g. [37, 180]) are treated as standard functionalities in robot control without increasing the level of autonomy.

Shared Autonomy or Semi-Autonomy

We use the term *shared autonomy* or *semi-autonomy* for all robotic systems that have some level of autonomy and where the human provides additional information or decisions, or completely guides the robot during execution. To visualize our understanding of the field of semi-autonomous systems, we modified a figure by Sheridan [191]. He related supervisory systems to their degree of automation and the task entropy that these systems are able to cope with (detailed below in the paragraph *supervisory control*). Figure 2.5a shows our understanding of the field of semi-autonomous systems ranging between industrial robots, that work almost completely automatic in predictable and known environments, systems that use direct teleoperation in unpredictable situations, and the ultimate robot that can handle unforeseen circumstances in a completely autonomous way. All the terms described in the following correspond to systems that are semi-autonomous.

Shared or Guided Control

Goodrich et al. [72] defined *shared control* as a level of autonomy where the robot uses an internal control loop to realize the continuously given commands by the human and alters them if necessary, e.g. by reacting to the environment in order to avoid obstacles. The term was also used in [7] to describe the level of autonomy that allowed blending of characteristics and superimposing inputs. Shortly, task execution is shared between direct control, local sensory feedback, and autonomy such that the human and the robot each control a subtask simultaneously [149]. The technique of virtual fixtures is also located in *shared control*. In this technique, the HMI displays additional feedback to the operator for assistance in fulfilling the task for example by constraining the motion along a certain path [31].

Traded Control

While in *shared control* human and robot work at the same time during the execution of a task, *traded control* was defined by [52] as a level of automation meaning "the consecutive assignment of subtask control to the human and machine". It means that the operator initiates a subtask or a behavior for the robot to perform and monitors its execution without further interference [72]. After the execution by the robot, the control is traded back to the human, who can initiate another subtask or use for example direct teleoperation to proceed. For the execution of behaviors, the human input reduces to set or trigger commands, and the robot replays pre-programmed behaviors, i.e. the fingers of a robotic hand form a certain shape after the human triggers the behavior. Another form of traded control is the setting of a goal point by the human, and then the robot performs motion planning to reach that goal point.

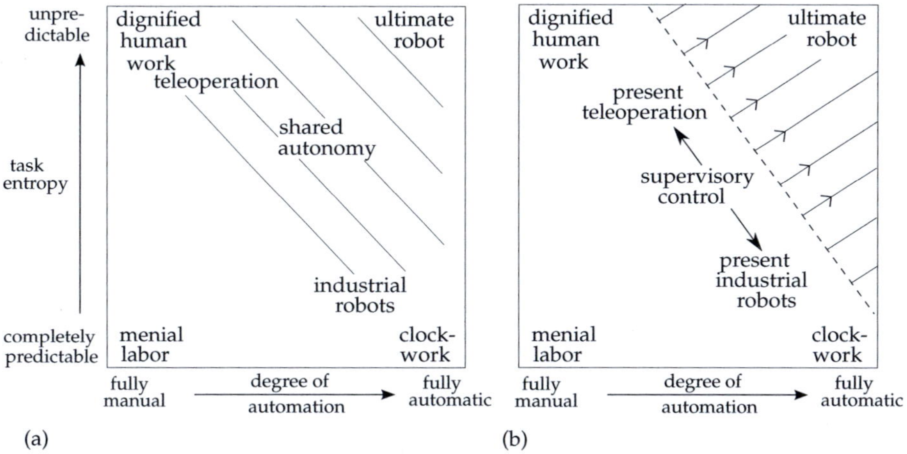

Figure 2.5.: (a) In this thesis, the term *shared autonomy* will be used covering all types of robotic system that combine some level of autonomy with input from the human. (b) Supervisory control relative to degree of automation and task predictability (Sheridan 1992, [191]).

Supervisory Control

The term *supervisory control* is used by Sheridan et al. [191] as a general term describing all robotic systems with some degree of autonomy. In their understanding, the human role in the system is not fixed but can vary from providing low-level input by changing control parameters to high-level input by only selecting the task goal. These robotic systems are classified in between directly teleoperated systems and industrial robots, as shown in Figure 2.5b. In contrast, *supervisory control* was used by Endsley et al. [57] as a level of automation where the human only monitors the autonomously working technical system (as described above) and gives high-level goals. We will also use the term in consensus with its linguistic usage, e.g. according to the Oxford dictionary, as synonym for observe and oversee. Hence, in our understanding of this classification of autonomy, the robot receives only symbolic, high-level goals from the operator (like "grasp the blue cup"). It autonomously plans its global behavior to reach the goal, for example by splitting it into known behaviors, and adapts the plan to the environment if necessary. The human solely monitors the execution, e.g. with a 2D video feed from the remote site. He does not need to be familiar with the subtask or behavior-based execution of the robot, but can focus on the task achievement. Upon intervention at failures, the mode of *supervisory control* can switch to other modes of shared autonomy like *shared* or *traded control* where lower-level control is possible.

Teleprogramming or Telesensor Programming

In telerobotic scenarios with a high time delay between master and slave, the method of *teleprogramming* or *telesensor programming* is mostly used [17, 64, 87]. There, the human interacts with a predictive display that shows the current (delayed) video feedback and a predicted state of the robot resulting from the current (undelayed) human input. The human commands are transferred to the remote site where the robot executes the programmed behavior according to the predicted model. Human input can either be on a higher language level [149] or continuous motion [35]. If the human interacts continuously with the simulation, behaviors, elementary motion, or symbolic commands can be extracted which are then commanded to the robot [64, 199, 200]. A more recent example is [169], where the Robonaut is operated with the use of a predictive display to trigger elementary behaviors. This approach is also used to create a natural user interface [117] where the human interacts with a virtual environment with gesture based commands like drag and drop to pick and place objects. These inputs trigger automatic object detection and parametrize a motion planner to provide a collision-free planning. As the grasp planner in this case can choose any kind of grasp that is valid, it might result in unpractical grasps for the task. In accordance with the Handbook of Robotics [149], we classify *telesensor programming* as a special form of supervisory control, as the human always interacts through an abstraction layer with the robot (the local simulation). Robot and human never share a simultaneous execution of the task nor is the control entirely traded to the human, but the robot always needs to execute the programmed behavior autonomously.

Collaborative Control

This term is often used to describe a team where human and robot act as peers and work together shoulder-to-shoulder in the same workspace [13, 90]. In telerobotic scenarios, *collaboration* emphasizes the information exchange between human operator and the robot in a supervisory setup. Hence, the interaction is not focused on the human to superintend the robot, but human and robot engage in a dialog to exchange ideas, ask questions, and to resolve inconsistencies [63, 72].

Mixed Initiative Control

As summarized by [5], the term *mixed initiative control* has many interpretations. Goodrich et al. [72] used the term for example as synonym for *collaborative control* focusing on the dialoge between human and robot, while [109] use it to avoid problems in *traded control* if the robot cannot monitor what parts of the task have been accomplished while the human was in control. Also, *mixed initiative control* can be used to describe a system where the control is not completely shifted to the robot during execution, but intermediate forms of *shared control* can be found [34].

Cooperative Control or Multi-user/agent Systems

In *cooperative control*, several operators and/or multiple remote robots work together to accomplish a task. For example, multiple operators cooperate to control a single robot via any form of shared autonomy [72], or several vehicles achieve a task in combination [143].

Autonomous Robot

The highest autonomy level refers to a fully *autonomous* robotic system where the human is neither needed to intervene nor informed about the cognitive reasoning of the robot. Technically, this is not a telerobotic system.

Others

Terms using *adaptive, sliding*, or *adjustable* autonomy refer to systems where the mode of shared autonomy can switch or can be adjusted during operation [110]. In comparison, the term *adaptive control* is understood as a control approach to automatically adapt the controller gains [9] and is not taken into account here. *Adaptive automation* is an automation that adapts to changes in the human operator or the environment, e.g. a system that adapts its display information based on the workload of the human or the task completion time [192]. Other terms like *direct agent control, manual control with intelligent assistance, strategic control*, or *partial autonomy* are also not considered.

The various, inconsistent uses of the terms introduced above, makes it difficult to estimate the work done in each sub-area. Nevertheless, we tried to visualize the broad field of shared autonomy systems by the number of results on `scholar.google.com`, accessed in January 2015, in Figure 2.6 (y-axis is in logarithmic scale). The search included each term in quotation marks plus the word "robot" to focus on robotic systems. Clearly, *supervisory control, cooperative control*, and *shared control* are mostly used in literature. Some terms have been extensively used in the past few years like *cooperative control, adjustable autonomy*, or *sliding autonomy*, while some terms like *teleprogramming* were more common in the 20th century. It is interesting, that very general expressions like *shared autonomy* or *semi-autonomy* have only few results compared to *supervisory* or *cooperative control*.

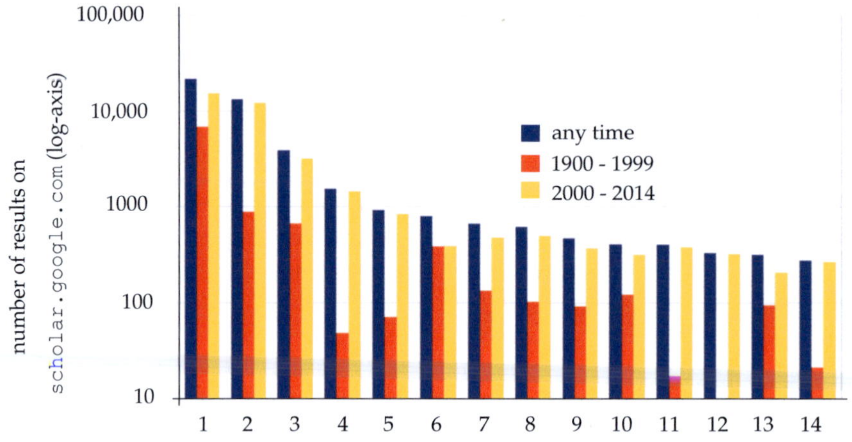

Figure 2.6.: Wording used for varying levels of autonomy and overall the number of results on scholar.google.com (January 2015): 1: supervisory control (21,600); 2: cooperative control (13,200); 3: shared control (3,890); 4: adjustable autonomy (1,530); 5: adaptive automation (917); 6: teleprogramming (799); 7: shared autonomy (666); 8: remote programming (616); 9: semi autonomy (469); 10: partial autonomy (405); 11: mixed initiative control (400); 12: sliding autonomy (331); 13: traded control (317); 14: adaptive autonomy (277)

2.2.2. State of the Art

This section reviews the state of the art in shared autonomy for telerobotic systems, where a robot is operated to perform manipulation tasks. We present the state of the art sorted in five levels of autonomy. After the categorization of the systems in these five levels, we discuss two additional important factors that have to be considered for shared autonomy systems in telerobotics: the dexterity of the remote robot, which determines the range of achievable tasks, and the immersion created by the HMI, which influences the mental workload of the human and his focus on the task.

The levels of autonomy range from *direct teleoperation* (Level 1), where the robot strictly follows the continuously given commands by the operator, to *supervisory control* (Level 5), where the operator only monitors the execution of the robot and commands high-level goals. The intermediate levels are sorted according to the amount of information sent to the robot and the autonomous ability to adapt to a new situation. Therefore, Level 2 is classified as *shared control* where the robot uses an internal control loop to realize and adapt continuously given commands. Level 3 and 4 are forms of *traded control* where either the robot or the human controls the execution of the task. The input of the human can be a continuous motion or is reduced to set or trigger commands. While in Level 3 the robot only replays preprogrammed behaviors or elementary functions, it requires general motion or grasp planners in Level 4 to execute its task. Of course, some systems incorporate combinations of different levels of autonomy.

State of the Art on Level 1, Direct Teleoperation

The focus of direct teleoperation lies mostly in creating a highly immersive system for the operator, and methods for bilateral teleoperation are usually analyzed, like in [37, 93, 209]. The robot strictly follows the continuously given commands by the operator (e.g. joint-to-joint coupling), but also optimizes the nullspace of the arm, and avoids singularities or joint limits [10, 112]. Level 1 is commonly used to control the end effector, for example in [37, 180, 198, 209], or as fallback strategy for the arm if other levels of autonomy fail [24, 48].

State of the Art on Level 2, Shared Control

For manipulation tasks, a common practice to increase the level of autonomy from Level 1 to Level 2 is to automatically optimize the posture of the robotic arm, while the human commands a continuous motion. Then, the Cartesian pose of the end effector of remote robot and HMI are continuously coupled, while the robot uses an internal control loop to realize and adapt the continuously given commands to the environment. This includes for example avoiding collisions with obstacles in the environment [125, 180]. Without influencing and modeling the motion to fulfill the task, avoidance of joint limits or of collisions with the environment can ensure a safe operation of the robot, i.e. as the motion simply stops before collision or the workspace limits. Then, the human can adjust his commands to move the robot to a safer region [180]. Also, the redundancy of the remote robot can be used to optimize its configuration for a certain goal, i.e. increase manipulability, while ensuring the task tracking by the slave [125]. Obstacle avoidance in combination with an optimal configuration of the robot can be modeled as repellent forces and displayed to the operator using haptic feedback [10].

The concept of assistance using force feedback leads to a popular method of support: virtual fixtures. These fixtures are constraints that guide the operator along a certain task-specific pathway or prevent him to enter forbidden regions, as reviewed in the survey papers [4, 31]. This restricts the motion of the manipulator into alignment with a task. For static virtual fixtures, this can result in a drastic reduction in flexibility of the telemanipulation system, as usually only a few simple tasks are considered [3]. Nevertheless, for specific applications like maintenance in a nuclear plant, a virtual fixture can guarantee task success, ensuring a certain orientation of a brush in combination with force feedback [48]. Another approach by [104] presented the use of "reflexes" that are applied during continuous commands of the operator. These reflexes are virtual fixtures preventing the end effector to leave the desired work area on the one side and on the other side attracting it towards a desired relative position to a goal. Overall, static virtual fixtures result in a decrease of variance in the desired trajectory and operator workload for the specified task [31]. Recently, approaches have been developed to dynamically adapt the virtual fixtures, for example [2, 24]. In [2], the task is first demonstrated and then the motion is divided into subtasks, where each subtask is represented as a straight line using Hidden Markov Models and Support Vector Machines. After learning the task, the method allows to handle deviations during execution in a flexible manner: when the classification of the current state in the task has a low match, the user has free control, otherwise the motion is constrained with the fixture. Haptic shared control is used by [24] to increase situation awareness of the operator with continuous haptic feedback, which is based on an optimal control action computed from additional information coming from the human, the task, and the environment. In order to train the operator for controlling a second order mass-spring-damper system, [151] analyzes the so-called active haptic cues. They compare static and dynamic virtual fixtures and find them to perform equally for a simple two dimensional movement to hit a target. In [131], the control of movements is shared between operator and autonomy, assuming a known task. The degree of sharing is realized with a task-space weighting matrix that enables smooth transitions.

Instead of influencing the commands from the human by virtual fixtures, they can be alternated by changing the mapping to the remote robot. This results for example in variable velocity scaling [55, 161]. The closer the end effector is located to a goal, the smaller the scaling, meaning that movements of the human result in small movements of the slave robot, thus enabling both fast approach motions far from the goal, and high precision motion close to the tool. An extension of [161] was provided by [225] where the movement scaling adapts according to a preprogrammed task sequence.

The direct mapping of every finger can clearly be enhanced by increasing the level of autonomy from Level 1 to Level 2. For example, the grasping forces are autonomously handled once a grasp is established [73]. Another approach by [106] splits the control of single fingers. For a pre-specified pinching task, one finger is directly controlled by the operator while the other stabilizes the pinching force automatically. Also, changing the mapping such that hand shapes are repoduced by the hand (instead of commanding single joint angles) increases task performance [121]. Introducing gestures as commanded motions allows to match the behavior of the manipulated object, instead of the finger motion by using a library of tasks [122, 206].

State of the Art on Level 3, Traded Control

Increasing the level of autonomy while still enabling the operator to command motions leads to traded control. Here, the level of autonomy is increased to Level 3 where the robot executes preprogrammed behaviors or elementary actions autonomously. These actions can be triggered, e.g. by a single mouse click, or automatically detected during the motion of the operator. Then, control is traded from the human to the autonomous execution of the robot. This reduces the demands on the immersion level of the human as the remote robot (at least partially) interprets and executes the commands autonomously. However in Level 3, autonomous re-planning due to changes in the environment is not required. Teleprogramming systems like [64, 169, 200] also use Level 3 during autonomous execution. A structured task environment allows the autonomous system to recognize the intention of the user and take completely over the control of the arm if a task can be automated [26]. During the autonomous execution of a subtask, the operator supervises and can stop the action if needed.

The main focus of [26] is to overcome time delay and a low bandwidth communication. The concept is proven to be valid by recognizing grasping tasks where a few grasp poses for each object are precomputed. An object is chosen to be grasped if the end effector is close enough and moves towards it. A similar approach is presented by [154], but the task to fulfill is learned by demonstration in advance, and encoded with Hidden Markov Models (HMMs). As soon as the system recognizes the human intent, it finishes the action autonomously and returns then the control back to the human who has to take over. The contact forces during maintenance tasks were used in [198] to encode constraint manipulation in HMMs. This allowed to create virtual fixtures that helped the operator to keep the task constraints. Although the time to complete the task did not change between assisted and non-assisted modes, operators acted more precisely and with less effort when assisted [198]. Instead of simply aborting the autonomous action, the human can intervene the executed path and move the robot around obstacles via direct teleoperation in [102]. There, he is also given the possibility to program the robot by selecting via points by pressing keys on a keyboard during a trajectory teaching. The trajectory is obtained by linear interpolation between these points, while the orientation of the end effector is constrained by the surface normal of the target object which is determined with several laser measurements. Whereas this method is a simple form of trajectory generation, it is yet very flexible due to the easy possibility for the human to intervene and reprogram [103]. Two assistance modes of visual guidance of an ideal trajectory and an auto pilot in a move-to-target task are compared in [118]. They confirm the results found in [57] that higher levels of autonomy (meaning the auto pilot) increases the routine performance and lowers the mental workload, but decreases performance in the case of failure. In the approach presented by [160], the operator teleoperates a robot with continuous motion with one hand and has the possibility to select objects with the other hand. After selection, the robotic arm uses visual servoing to reach a preprogrammed pose of the arm relative to the target object to grasp it. This overcomes the problem for visual servoing to continuously see the target object with the camera as the operator can first approach the object and let the fine motion for grasping be executed by the robotic arm.

Level 3 of autonomy is commonly used for the end effectors as most systems use preprogrammed grasps, meaning preprogrammed finger configurations, to grasp an object. For end effectors with one DoF, the open and close motion is usually triggered (e.g. [43, 116, 162, 220]), while for more complex end effectors several grasping shapes can be realized [89, 93, 112, 113, 135, 223]. Similarly, preprogrammed manipulation primitives can be triggered with a low DoF input device [136].

State of the Art on Level 4, Traded Control

The robot plans its local behavior according to the triggered command and can adapt it to the environment (e.g. on a set command of a goal point, the robot issues a motion planner to reach the goal and to avoid dynamic obstacles during execution).

When Level 4 of autonomy is reached for the arm, usually the immersion level of the operator involves 2D visual feedback on a computer monitor (without haptic feedback) as the main interactions are triggering commands instead of continuous motion. Nevertheless, most systems still allow for setting low level joint angles as a fallback strategy if autonomy fails. The view on the remote scene, the commands, and motions are issued with a computer mouse. The

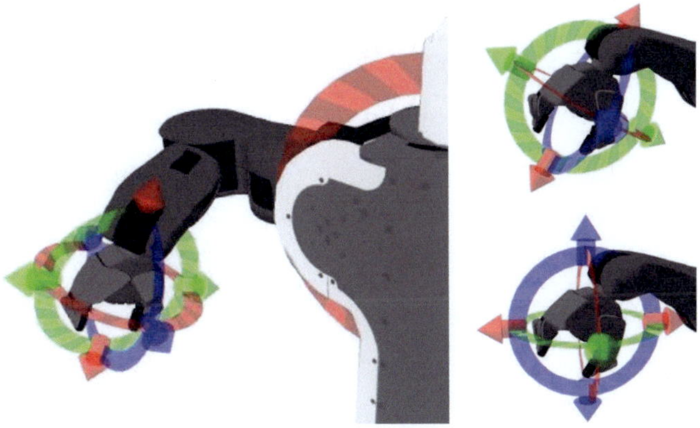

Figure 2.7.: The interactive markers shown are used to move the pose of the end effector [116]. Left image: "Gripper control (6D) and shoulder ring (1D). Right images show a gripper-aligned (top) and world-aligned (bottom) control."

majority of work in this level of autonomy of the arm is contributed either by teams working with the PR2 [65] or in the context of the DARPA robotics challenge (DRC) [210].

In combination with the development of the PR2, Willow Garage also developed the Robot Operating System ROS. Its standard viewer is called RViz [153] and it is widely used in the robotics community. It shows an RGB view of a camera and a rendered 3D scene, e.g. obtained from a pointcloud. Inside the visualization, the concept of interactive markers is used to select an end effector or an object and move it along a specified axis with the computer mouse (Figure 2.7). This concept is useful to quickly correct the position of a robotic arm, but can be tedious when it comes to commanding large motions or moving every fingertip of a multi-finger hand. This means that the introduced autonomy mainly overcomes the limited input capabilities, which for instance leads to performance improvements of a mobile manipulation system in cluttered environments [43], or to higher efficiency in grasping tasks [116]. Grasping was performed with a two-finger gripper that closes on trigger commands. The grasp pose of the end effector relative to the object was either precomputed and retrieved from a database or generated online from a point cloud. Then, the motion of the arm to that target pose is autonomously planned [116]. This concept was further developed by [220] where the user refines context information for the task such that a suitable grasp for a task can be chosen.

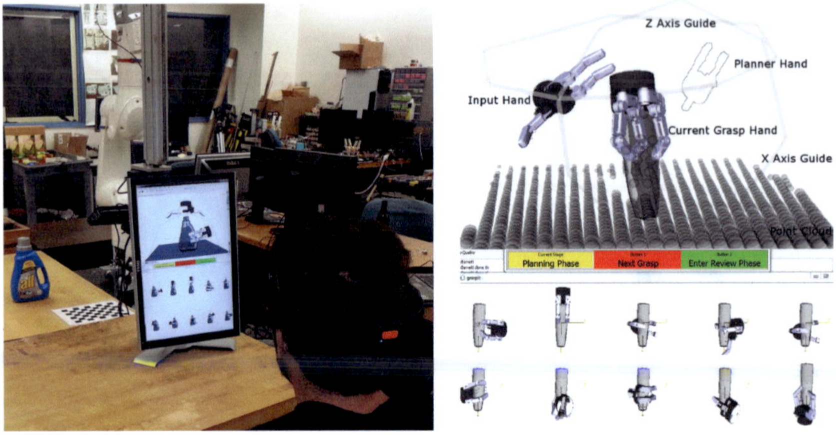

(a) (b)

Figure 2.8.: Images adapted from [217]. (a) User with a brain computer interface controlling the system. (b) The grasp planning interface where the human can select grasps and adjust the opening angle online.

In the DRC, several tasks have to be completed with a humanoid robot like walking, climbing a ladder, turning a valve, or driving a car. Although direct teleoperation is allowed, it is disturbed by communication swapping between a high quality (1 Mbps, 50 ms one way latency) and a low quality (100 Kbps, 500 ms one way latency) link. This motivated the participating teams to investigate in various forms of shared autonomy where for example balancing is performed autonomously while the step planning is carried out by a human operator [98]. In general, the tasks required whole body planning and multi-contact planning for robust grasps rather than fine manipulation of objects. Therefore, almost all teams use a form of drag and drop commands to position the end effector. Many teams also used the viewer RViz and the concept of interactive markers to control the robot arm on this lowest level. Given the goal point, the motion of the arm is solved via inverse kinematics [60, 98] or a motion planner is employed [60, 130, 162]. Another strategy is based on motion primitives that can be adapted to the current environment [190]. Despite amazing performance of some teams, they reported that the workload of a single operator was too high to be handled. Therefore, teams used several operators with dedicated tasks, e.g. a perception operator [60] or a GUI operator [130], to control the robot. Overall, the concept of shared autonomy proved to be very effective for the DRC if principles of effective human-machine teamwork were considered during the design phase as pointed out by [98]. They argue against traded control, where the control is completely shifted to the robot, and rather decided to use shared autonomy systems that combine the capabilities of human and robot. Also, multiple ways for executing a task are important to guarantee a high success rate. In the context of the DRC, teams did not describe in detail how they handled the planning for the fingers of the end effector. We assume that they use pre-specified grasps or follow a simple close-joints strategy (Level 3) [130, 162, 190]. Only Team MIT [60] states that they could issue an optimization-based grasp planner solving for feasible grasps that satisfied force closure constraints. These grasps were planned for the Robotiq gripper with 4 DoF, but it is unclear if they actually used the planner during the trials [60].

Besides these works, a shared autonomy system that uses grasp planning, and thus reaches Level 4 of autonomy for the end effector, is presented by [217]. It is developed in the context of assistive robotics where the operator commands grasps for the three finger Barrett hand using a brain computer interface (four facial gestures) to select and execute grasps (Figure 2.8a). The operator can browse through a list of precomputed grasps from a database and select the most suitable. The selected grasp can either be executed or used as initial guess to rerun a grasp planner in order to find grasps similar to the selected one. The mean time to grasp an object (from the end of the object detection to the beginning of the motion of the robotic arm) ranges from 53 s to 151 s.

State of the Art on Level 5, Supervisory Control

The robot receives only high-level goals from the operator, can autonomously plan its global behavior to reach the goal and adapt it to the environment if necessary. A concept for such a shared autonomy system was developed by [59] for a health care application in hospitals, although a detailed description of the capabilities of the supervised robot arm is missing. More recently, several systems have been developed for using tablet computers as interface to supervise the robot, for example [23, 145]. While [145] provides the user with multiple skills that he can sequence for a task, [23] uses an object-centered approach. The operator can select feasible actions for each object, which are an autonomously chosen selection of all possible actions assigned to the object, dependent on the current situation. It is important to note that these systems use Level 3 for the end effector, i.e. preprogrammed grasps.

Discussion and Summary

The presented classification of shared autonomy systems in telerobotics for grasping and manipulation according to five levels of autonomy does not explicitely consider the range of tasks that a system can achieve nor the immersion that an HMI can create. Therefore, we visually summarize the previous work in Figure 2.9, where the dexterity of the remote robot is represented on the x-axis as number of DoF involved in the manipulation task, and the y-axis displays the level of immersion. The level of autonomy is considered separately for the arm and for the end effector, and each one is represented with a half-circle (arm in blue with a colored area, end effector in orange, see legend at the bottom of the figure). The fuller the half circle, the higher the level of autonomy.

As all parts of the task are executed by a robot, its manipulation capability directly determines the achievable complexity of tasks for dexterous and fine manipulation. We assume that with an increasing number of DoF, the robot can fulfill more complex tasks. In Figure 2.9, we consider only the DoF of the system involved in a single-arm manipulation task, comprising one arm, one hand, and the head. Therefore, we treat bimanual tasks as the extension of single-arm manipulation. Although previous discussion also included humanoid systems, and even though the legs might be important for a humanoid during the task to stabilize the robot, these additional DoF are not taken into account.

The immersion possibly created by an HMI is rated highest if all modalities, that the human would experience during execution of a task with his own body, can be provided. For manipulation with one arm, this implies commanding of motion in all DoF of the human hand base, the fingers, and the head. Furthermore, haptic feedback for fingers and arm, and stereo vision with changing viewpoint are considered. Additionally, the term "audio" summarizes all types of audio feedback (like mono, stereo, and 3D feedback). In comparision, we distinguish between 2D and 3D visual feedback. A 2D visual feedback on a screen with a fixed viewpoint without the possibility of commanding continuous arm or finger motions to the robot is rated lowest in terms of immersion.

In order to simplify the reading of Figure 2.9, we summarize all cited state of the art works in Table 2.3. The last two columns give the quadrant of the system in the figure, i.e. Aarno et al., 2005, [2] can be found at x position 10 and y-quadrant E. The table is sorted according to the citation number.

In summary, there are many concepts and systems that use shared autonomy for grasping and manipulation. Most of them focus on controlling the robotic arm and end effectors are either neglected (e.g. [81, 151]) or have only a few DoF, thus simplifying the control and planning (e.g. [43, 116]). The most popular approaches are either to trigger preprogrammed grasps (e.g. [89, 93, 113, 135, 223]) which heavily reduces the capabilities of a dexterous end effector, or to manually control the hand with a dataglove (e.g. [180, 209]). Approaches that split the grasping process in teleoperated motion and autonomous stabilization of grasp forces on the object (e.g. [73]), as well as methods that use manipulation primitives (e.g. [136]) are limited to a few specified tasks. In general, it is difficult to port concepts for teleoperation and shared autonomy for the arm to the hand. Although the use of interactive markers as shown in Figure 2.7 might be sufficient to manually move the pose of an end effector on a 2D screen, it can be very tedious if not impossible to position every finger of a dexterous hand on an object and command it to squeeze it. Additionally, some fingers might be invisible within the available camera viewpoints during the grasping process if the object occludes them. Most importantly, a grasp has in general the goal to hold the object stable and to withstand external disturbance forces. This makes it difficult to judge the quality of a grasp based on visual information only, as it is shown in [217]. To sum up, grasp planning for multi-finger hands is seldomly used in shared autonomy systems, and never in telepresence scenarios where the operator is highly immersed in the remote environment.

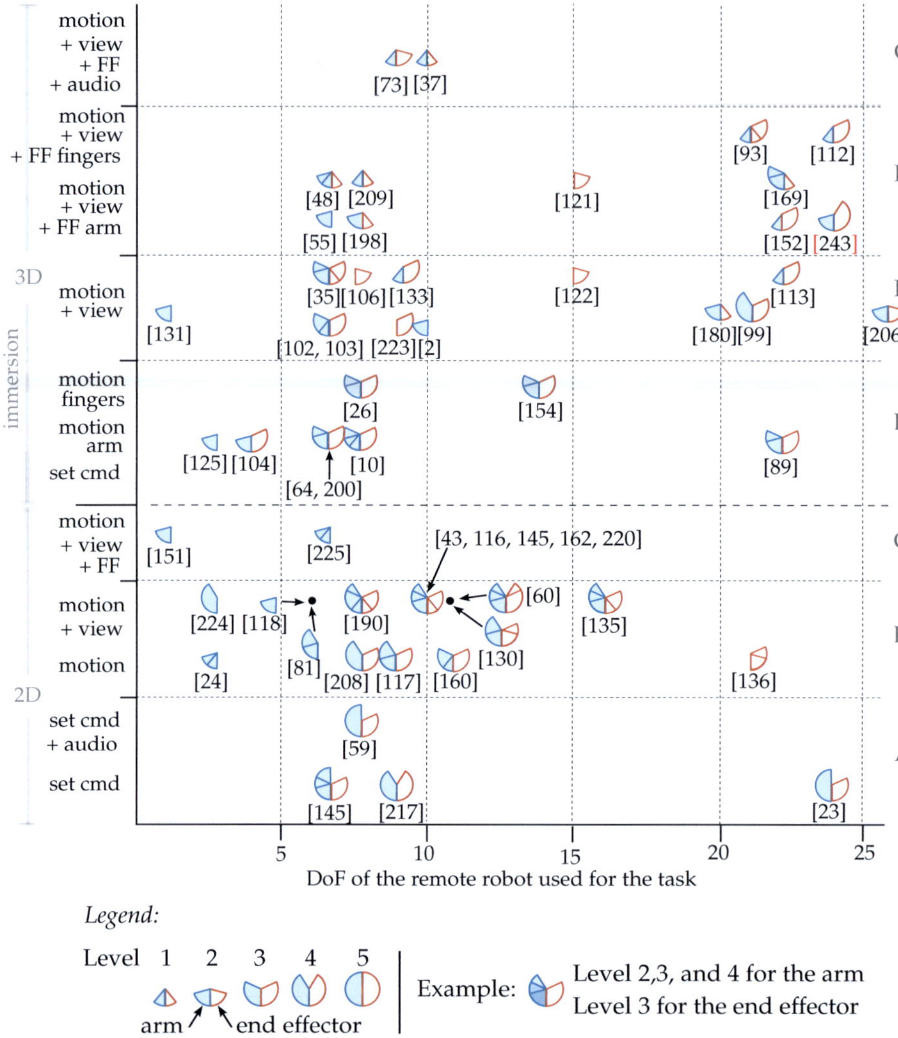

Figure 2.9.: Summary of state of the art shared autonomy systems for grasping and manipulation. For each system, the level of autonomy is depicted as a half circle for the arm (left, blue semicircle) and another one for the hand (right, orange semicircle), and a "full" half circle corresponds to Level 5. If several levels of autonomy are realized, the corresponding half circle shows multiple levels. The systems are sorted according to the number of degrees of freedom involved in the manipulation task (usually, one arm, a hand, and the head) and the level of immersion. We assume immersion increases from 2D to 3D view; from setting trigger commands (set cmd) to commanding motions and influence on the viewpoint on the remote scene (motion+view). Then, force feedback (FF) and audio feedback increases immersion even more.

Table 2.3.: Authors and publications in Figure 2.9

Authors	Year	Citation number	Quadrant at y-axis	Quadrant at x-axis
Aarno et al.	2005	[2]	E	10
Backes and Tso	1990	[10]	D	5–10
Birkenkampf et al.	2014	[23]	A	20–25
Boessenkool et al.	2013	[24]	B	1–5
Bohren et al.	2013	[26]	D	5–10
Brunner et al.	1993	[35]	E	5–10
Buss et al.	2010	[37]	G	10
Ciocarlie et al.	2012	[43]	B	10
Desbats et al.	2006	[48]	F	5–10
Dubey et al.	2011	[55]	F	5–10
Ettelt et al.	1998	[59]	A	5–10
Fallon et al.	2014	[60]	B	10–15
Funda and Paul	1990	[64]	D	5–10
Griffin et al.	2005	[73]	G	5–10
Hauser	2013	[81]	B	5–10
Hochberg et al.	2012	[89]	D	20–25
Hu et al.	2005	[93]	F	20–25
Johnston and Rabe	2006	[99]	E	20–25
Khokar et al.	2011	[102]	E	5–10
Khokar et al.	2010	[103]	E	5–10
Kim et al.	2006	[104]	D	1–5
Kobayashi et al.	2012	[106]	E	5–10
Kremer et al.	2009	[112]	F	20–25
Krueger and Schiele	2012	[113]	E	20–25
Leeper et al.	2012	[116]	B	10
Levine et al.	2014	[117]	B	5–10
Li et al.	2014	[118]	B	5–10
Lii et al.	2010	[121]	F	15
Lii et al.	2012	[122]	E	15
Liu and Chopra	2011	[125]	D	1–5
Mainprice et al.	2014	[130]	B	10–15
Malysz and Sirouspour	2012	[131]	E	1–5
Markovic et al.	2014	[133]	E	5–10
Mast et al.	2012	[135]	B	15–20

Table 2.3.: Authors and publications in Figure 2.9

Authors	Year	Citation number	Quadrant at y-axis	Quadrant at x-axis
Michelman and Allen	1994	[136]	B	20–25
Muszynski et al.	2012	[145]	B	10
O'Malley et al.	2006	[151]	C	1–5
O'Malley and Ambrose	2003	[152]	F	20–25
Padoy and Hager	2011	[154]	D	10–15
Perez Quintero et al.	2014	[160]	B	10–15
Phillips-Grafflin et al.	2014	[162]	B	10
Rehnmark et al.	2005	[169]	F	20–25
Rosell et al.	2013	[180]	E	20
Settimi et al.	2014	[190]	B	5–10
Stefanov et al.	2013	[198]	F	5–10
Stein et al.	1995	[200]	D	5–10
Toh et al.	2012	[206]	E	25–30
Tsui et al.	2011	[208]	B	5–10
Turner et al.	2000	[209]	F	5–10
Weisz et al.	2013	[217]	A	5–10
Witzig et al.	2013	[220]	B	10
Yoshimura and Ozawa	2012	[223]	E	5–10
You and Hauser	2011	[224]	B	1–5
Yu et al.	2003	[225]	C	5–10
Hertkorn et al.	2013	[243]	F	20–25

2.3. The Shared Grasping Approach: a Combination of Telepresence and Grasp Planning

The summary of related works, presented in the previous section, revealed a lack of shared autonomy systems that enable grasping and manipulation with a dexterous hand and a high level of autonomy (Level 4 for the hand), while simultaneously providing an intuitive interface for the human operator. This is mainly due to the hardware challenges that are imposed by the human hand both as input device as well as output device on the side of the remote robot.

In comparison, this thesis presents an approach for combining grasp planning with a telepresence system and hence, developing a system that provides a high level of autonomy of a dexterous hand with an intuitive interface for the human. Our system can be found in Figure 2.9 in the upper right corner F, 20–25 [243]. This section presents the background for the specific approach developed in this thesis.

We consider the interface to the operator as crucial in a teleoperation system because not only the autonomy of the remote robot but also the interaction with the human determines the performance of the telerobotic system. On the one hand, empirical studies show an increase of performance using a higher degree of autonomy while decreasing the workload of the human operator [100]. On the other hand, for high levels of autonomy, a weaker sense of telepresence and lower performance in situations that require failure recovery are reported, resulting from "out-of-the-loop" phenomena [58]. This can happen when the human is not completely immersed in the remote environment and his situational awareness is too low to detect system failures. According to [57, 58], the human is generally ill-suited for a long-term monitoring of automated processes. Sometimes, an increase in autonomy of the robot can even lead to more complex interaction for the human operator [11]. Therefore, we choose a moderate level of autonomy which is more appropriate especially in complex environments; this is obtained by automating some parts of the tasks, while others are left under human control. For example, including additional visual information might increase system transparency, mitigating the effect of mapping inconsistencies and unnatural interaction modes. However, this might lead to problems like visual cluttering, which results in a higher cognitive load for the operator and a slower execution of tasks. These results stress the importance of a careful design of the autonomy level of a remote robot in a telerobotic setup. The human should be able to use the advancements on the robotic side effectively while still perceiving sufficient and appropriate control over the robot. Then, the abilities of a human operator in terms of judgement and reasoning can be combined with the precision and repeatability of autonomous robots. Like the factors influencing the sense of telepresence, the design of the interface and the needed control over the robot heavily depend on the task and the experience of the operator.

We first analyze a control scheme of human prehension in order to deduce requirements under which the later described assistance for grasping was developed. Due to the importance of keeping the human in the loop and focused on the task, we want to preserve the high level of immersion of the human

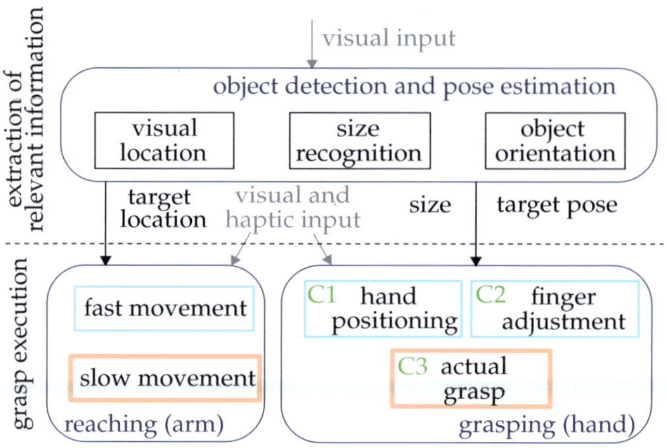

Figure 2.10.: Human control scheme for prehension tasks. It can be divided in two main parts: extraction of relevant task information from the visual scene (on the top), and execution (on the bottom). The execution can be separated into reaching (on the left) and grasping (on the right) which are synchronized motions. Positioning of the hand and finger adjustement correlates to a high speed of the reaching motion (blue squares) while the actual grasp is executed during slow movement of the arm (orange squares) [129].

operator in telepresence systems. Hence, we combine requirements from both human prehension and teleoperation, and derive a new framework for assistance.

Human hands and the process of grasping and manipulation are active research areas. In various fields like psychology, neuroscience, or mathematics, researchers try to understand human prehension by performing empirical studies or by building computational models in order to verify their conceptual models. It is not clear yet, how humans perform grasping and how they control their hands, and several contradictory findings have been presented, summarized for the neuroscience of grasping for example in [40]. Nevertheless, the main phases of natural prehension behavior could be identified: Figure 2.10 shows the model of Arbib (according to [129]) dividing prehension into two main sequential parts: extraction of relevant task information from the visual scene, and grasp execution. Some conceptual models describe prehension as a distributed action, as proposed by Arbib and also Paillard, in contrast to hierarchical models presented by Brooks, Keele and Green [129]. However, they all share the description of prehension as distributed motor schemes that are synchronized during the execution.

The model of Arbib (Figure 2.10), in compliance with other traditional approaches, identifies two phases during approaching an object: a high and a low velocity phase of the arm [96]. During the high velocity phase the fingers are preshaped to adapt to the object geometry; fingers are closed during the slow movement of the arm to actually perform the grasp. Prior to the movement, the grasp is planned by selecting a grasp strategy and planning a hand location and orientation. This leads to suitable positions for the fingers on the object's surface [196]. More recent results confirm that a sub-sequential execution can be separated into reaching and grasping phases [101]. The movement is influenced by several aspects like the physical properties of the object [129] or the visibility of contact points [196]. A recent result even postulates that human grasping is guided by the finger that aims toward the visible contact point on the object [215].

According to the presented grasping motor scheme of the human, three main challenges for executing a grasp in teleoperation systems are identified (Figure 2.10):

C1 - hand positioning
: Setting up a position and orientation of the hand relative to the object.

C2 - finger adjustment
: Finding contact points and the corresponding finger joint configurations for a stable grasp.

C3 - actual grasp
: Moving into contact with all fingers, applying suitable contact forces, and ensuring grasp stability.

From a telerobotics perspective, all three requirements can be assigned to properties of the system. C1 and C2 are highly influenced by the chosen mapping between human hand/arm and robot: C1 directly relates to the reachability of the grasps planned in C2. Increased difficulty (also affecting C3) comes with higher differences in kinematic and dynamic properties between robotic and human hand. The process of finding suitable grasping points on an object surface and the corresponding hand pose relative to the object needs to be adjusted according to the robotic end effector used in the system and to the object. This procedure might not be intuitively understandable for an unexperienced operator using direct mapping. If a limited number of pre-programmed hand postures are used as grasps, the hand positioning needs to be learned (or guessed) by the human operator. This is particularly crucial if the robotic end effector has a large number of degrees of freedom, but is not anthropomorphic [21]. Performing the actual grasp (C3) with some visual occlusion and without sufficient haptic finger feedback makes it almost impossible for the human to command the desired grasping forces in direct teleoperation. This results in undesired behaviors like grasp failure, toppling or dropping the object.

Most shared autonomy systems developed so far using complex end effectors neglect C1 and C2 and just approach the problem of performing an actual grasp (C3), e.g. by replacing the control of the finger movement by the user with a trigger for a preprogrammed autonomous grasp, as presented in the previous section. In contrast, assistance for grasping and manipulation needs to tackle all described challenges, while enabling the human to perform the natural reaching behavior, and keeping a high level of immersion by the active participation in the prehension process. To realize such an assistance, two aspects have to be considered: Aspect Handling and Aspect Interface.

Aspect Handling

The human operator needs to be able to use and apply his task knowledge, as well as his ability to solve problems in unexpected situations. Therefore, the system should be usable without expert knowledge about the robotic system. This aspect is highly influenced by the grasp direction, i.e. from which direction the hand should grasp the object, and by the part of the object surface that the fingers touch to grasp the object. The task knowledge of the human can be incorporated by choosing grasps based on the relative pose of the hand with respect to the object. This requires a grasp planning that is based on a hand pose relative to the object and that can cope with the speed of motion of the operator (C2). The outcome of the grasp planning phase are contact points for the fingers that should ensure a stable grasp and guarantee that the object can be manipulated without exceeding the torque limits of the fingers. This concept overcomes the limitation of pre-programmed grasps, by incorporating task knowledge and the correction of errors, and of a static mapping, by planning and hence guaranteeing grasp stability. Like in previous shared autonomy approaches, the actual grasp (C3) is executed autonomously. The closing of the robotic fingers is triggered by the human operator. It is important to note that the operator should be able to choose a different grasp at any time using the simple trigger in combination with re-positioning the hand relative to the object. Additionally, an assistance for the hand pose (C1) has to be provided indicating the feasibility of grasps to support the human in finding a hand pose relative to the object that most likely leads to a stable grasp.

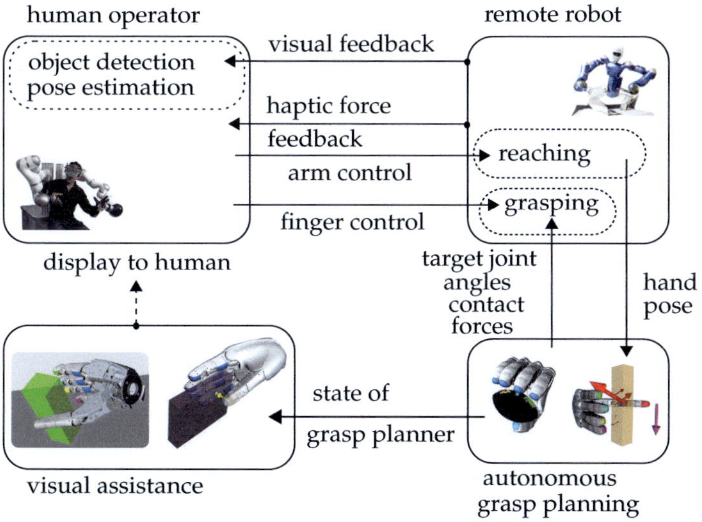

Figure 2.11.: Schematic overview of the shared autonomy system. While the human still processes the visual information to detect the object and estimate its pose, the execution of the prehension task is commanded by the human but performed by the remote robot.

Aspect Interface

Additional information provided by the autonomous functions should be appropriately displayed to the human operator. The field of augmented reality has shown that the display of additional information given by autonomous functions is helpful for humans to learn and fulfill tasks, e.g. for mechanics who assemble parts [83]. Thus, providing additional information for the human operator during the manipulation task can be beneficial. For example, the information if a feasible grasp can be executed in the given relative hand pose shoud be visualized to avoid that the human operator has to reason about the robotic capabilities. Considering the complexity of the overall telepresence system, simple visualization techniques, like changing the color of objects, seem desirable to avoid increasing the workload of the operator unnecessarily.

In the context of this work, a framework for a shared autonomy system was derived based on the reasoning above. Its structure is shown in Figure 2.11, where the dotted boxes correspond to the parts of the human control scheme for prehension (Figure 2.10), and the arrows show the links between the parts of the telerobotic system. While the human still processes the visual information to detect the object and estimate its pose, the execution of the prehension task is commanded by the human but performed by the remote robot. The next two chapters present the algorithms enabling assistance for C1 and C2 by using autonomous grasp planning. The calculation of the actual grasp based on the given hand pose is described in Chapter 3 and the method to guide the hand pose in Chapter 4. Both chapters include the analysis of the developed algorithms. Chapter 5 describes then the implementation of the concept on two hardware platforms at DLR and the development of the visual assistance. As an evaluation of a telerobotic system has to consider the human operator, we examine all developed assistance methods with appropriate user studies in Chapter 5.

3
Online Grasp Planning

This chapter describes the algorithms developed to assist grasping in teleoperation systems. The concept of assistance introduced in the previous chapter leads to the following requirements for the grasp planning:

General: As tasks are completed with the human in the loop, a major goal is to sustain his task knowledge and therefore, the grasp planner should restrict the possibilities to grasp an object as little as possible. Grasps on the overall surface of the object, with varying numbers of fingers, and without training should be possible.

Robust: Reaching the exact contact points using a real robotic system can hardly be guaranteed due to different sources of uncertainty, including errors in the object location, noisy sensors, or inaccurate models of the robot kinematics and dynamics. Hence, the computed grasps should be robust to be able to deal with these uncertainties.

Realistic: The grasp planner needs to take into account the kinematic and dynamic behavior of the robotic hand to ensure that the computed grasps can be actually executed by the hand.

Multi-finger hand: The concepts are applied to a system using a multi-finger hand. This type of end effector provides a large flexibility to handle different tasks, including grasping objects of different shapes and sizes.

Fast: The grasp planner needs to be able to cope with the speed of motion of the operator as he controls the hand pose.

This chapter includes content of previous publications [245], [251], and [252]. It starts with a description of related works (Section 3.1) and introduces then the preliminaries for grasp planning where the underlying assumptions are presented (Section 3.2). The developed grasp planning is based on the computation of independent contact regions [172], shortly summarized in Section 3.3. The algorithm was extended with a reachability and a feasibility check, i.e. the joint and the torque limits of the robotic hand are now considered during grasp planning. Both extensions rely on a representation of the workspace of the robotic hand presented in Section 3.4. Then, the algorithms for realistic grasp planning are introduced and evaluated in Section 3.5 and Section 3.6.

3.1. Related Works

Grasping an object implies finding contact points, contact forces, and a suitable hand configuration (position and orientation of the hand and its joint angles) such that the computed contact points and forces can be reached and applied by the robotic hand. In addition, the grasp is selected such that a good performance in the real environment can be guaranteed, for instance avoiding collisions with obstacles and resisting perturbations on the object. Several approaches to grasping have emerged ranging from purely analytical concepts in 2D [127] to machine learning approaches without the use of analytic grasp validation [184], summarized in [25]. Additionally, aspects like the number of fingers of the hand or the amount of prior object knowledge differ between the approaches.

In this thesis, we use an analytic approach as it provides quantifiable evaluation of the grasp properties such as equilibrium and stability, and also allows to find grasps in situations that the system was not trained for. These measures are especially crucial when it comes to grasping with multi-finger hands where a simple push with one finger or an unsynchronized touch of the object surface can be sufficient to lead to grasp failure of an otherwise stable grasp. Also, simplified "close-fingers" strategies like with two-finger grippers do not meet the capabilities of the high dimensionality of the configuration space of a multi-finger hand. However, it is important to note that analytic approaches rely on the correct modeling (and execution) of the grasp [22] and in order to make the problem tractable, assumptions like rigid-body modeling, simplified punctual contact models, and known objects are used. This emphasizes the need of taking into account uncertainties in the grasp planning. In analytic approaches, the three problems in grasping of finding contact point, contact forces, and a hand configuration are traditionally tackled as independent problems which has a major drawback: the contact points initially computed might not be reachable by a particular hand and the required contact forces might not be applicable by the robotic fingers. This could force the planner to iterate between finding contact points on the object [165, 235], computing the corresponding hand configuration, for example using heuristics [29] or optimization [179], and check for the contact forces [36, 76].

Approaches that lead to contact points that are reachable with a particular hand configuration use for example simulation tools like [49, 139, 211] and store the resulting grasps in databases [71, 82]. During runtime, a grasp is chosen according to the current environment or a given aim of the task. Although this approach is appealing due to the cheap computation of grasps, it is not easily adaptable for the use in our shared autonomy concept. As the human commands the hand pose relative to the object, the grasp database would need to be very dense despite the high dimensionality of the hand configuration space. An approach that iteratively utilizes a wrist position and orientation provided by the user to compute force closure grasps on the object uses eigengrasps [44]. They reduce the high dimensionality of the hand configuration space by obtaining hand motions with a principal component analysis of the full configuration space. However, the computational times do not allow the application of such a system in real-time applications. Another approach analyses the finger workspace offline to obtain all the possible contact points between a particular hand and object, for a fixed position and orientation of the hand [222]. Fast online generation of potential grasp candidates can be obtained by using a heuristic to quickly discard unreachable grasp configurations, although the heuristic is usually suited for a particular hand [28].

The above mentioned approaches neglect the contact forces, which are considered as a separated problem that is solved via grasp force optimization and grasp control (where, on the other hand, contact points are assumed to be known). Finding contact forces has been in general tackled avoiding the linearization of the friction cone constraints. Instead, they are formulated as the positive definiteness of suitably defined matrices [36], or reformulated as a convex optimization problem using linear matrix inequalities [76]. These formulations of the friction cones allow a general description and solution to the grasp force optimization problem [32, 123, 187]. Multi-objective optimization can be used to avoid a solution of the grasp force optimization that leads to contact forces being on the boundary of the friction cone, as that is error-prone [75]. Another technique solves the force optimization problem in the wrench space with a ray-shooting approach [233]. Rosales et al. [178] find a suitable hand configuration for known contact points using a model of compliance. They use constrained optimization to tackle contact reachability, force controllability, and object restraint simultaneously. While there are many data-driven approaches that consider contact forces without explicitly planning them [25], approaches for an analytic computation of contact points and forces simultaneously are usually based on optimization techniques or considered implicitly with a grasp quality measure. El-Khoury et al. [56] formulate for example a constrained optimization problem to obtain reachable grasps that are optimal in terms of a force related quality measure. They model the object as superquadric and manually select a region of interest on the object. The friction cone is linearized and they assume that the overall maximum applicable force of the hand is bounded.

To evaluate a grasp, several grasp quality measures have been proposed [174]. For the most common one, the test for force closure can be used to compute the maximum magnitude ε of a disturbance wrench that can act on the object in any direction without breaking the grasp closure. This ε-metric is either obtained by assuming a common power source for all the fingers [61], which simplifies the computation, or by having an independent power source for each finger [28, 231]. Other grasp quality measures were developed using physically motivated task wrench spaces [30] or taking into account the geometry of the object [202]. Most importantly, all these grasp quality measures are computed in a wrench space based on normalized contact forces, i.e. the maximum contact force per finger is smaller or equal to one. This neglects that force closure grasps need to be transformed to stable grasps by considering the realistic (not normalized) contact forces [147]. Additionally, this procedure only works if the single contact point and the optimized contact forces can be encountered accurately on the real object, which assumes that the robotic fingers are strong enough to counteract the expected disturbance wrench, i.e. the torque limits of the fingers are not exceeded. The hand configuration plays also an important role as the maximum force that a robotic finger can apply in a certain direction changes drastically within its workspace. Therefore, the success in the real world of the drastically within its workspace. Therefore, the success in the real world of the contact points planned with the ε-metric highly depends on these contact forces (and the hand configuration). Approaches to evaluate grasps with a physics-based simulation focus on finding stable grasps, exploring the fact that force closure is a sufficient but not necessary condition for stability [92]. Also, force closure can be tested for stability by perturbing the object pose, which can easily be done in simulation [105]. On the other hand, there is usually a simplified strategy to close the fingers in simulation and also the contact force distribution is not taken into account, which is crucial for successful grasps with multi-finger hands.

Among the analytic approaches, there is the concept of independent contact regions (ICRs) that is used in this thesis to provide semi-autonomous grasp planning for the human operator. ICRs provide a contact region instead of a single contact point that satisfies the force closure condition, and therefore they provide robustness to uncertainties in the grasping pipeline [148]. Starting from an initial force closure grasp, the concept can be used to geometrically construct families of grasps [164] or to compute contact regions on discretized objects [114, 172] using both frictionless and frictional contact models. The friction cone is always discretized, its magnitude normalized, and a common power source for all fingers is assumed. The approach presented in [172] creates a conservative search space that leads to a quick computation of the regions around an initial force closure grasp. [114] considers a more complete condition for computing the contact regions that includes the linear combination of the primitive wrenches for a given contact point, but is computationally very expensive due to an inclusion test that repetitively solves a linear programming problem. Actual executions of grasps involve different sources of uncertainty that influence the computation of the contact regions; for instance, uncertainties in the location of the contact points and their normal direction can be considered by reducing the problem to the selection of a suitable (and conservative) friction coefficient for the computation [170]. Further extensions of the concept involve the computation of ICRs for vision-acquired object models [41], and for planning of in-hand manipulation strategies such as regrasp [173] and rolling [97]. All of these approaches do not consider the hand kinematics, but assume only one contact point per finger. Nevertheless, the concept of ICRs can be used to compute force closure grasps that are robust to uncertainties in position of the fingers and/or the object.

The methods described in this chapter are a reformulation of the ICR algorithm in [172] and extend the ICR approach to synthesize reachable and feasible fingertip (precision) grasps. They are all based on a given hand pose relative to the object as that is provided by the human operator. The extended algorithms make use of the workspace of the robotic hand (Section 3.4). In Section 3.5, the workspace guarantees the computation of reachable contact points while the corresponding finger configurations are obtained that are needed to touch the object surface. The evaluation shows that the computation time of the method is fast enough to cope with the movement speed of the human operator. To overcome the limitation of normalized contact forces and to transform the planned force closure grasps into stable grasps, the contact forces that the finger can apply on the object surface depending on the current hand configuration are taken into account in Section 3.6. The method uses a ray-shooting approach (an extension of [231]) to determine the maximal external disturbance wrench that can act on the object while simultaneously shaping the wrench space according to the available contact forces taking into account the torque limits of the fingers. This allows for a realistic evaluation of the grasp.

3.2. Preliminaries

In this work, the object surface is represented by a cloud $\mathcal{P}$ of N points. The points are specified by position vectors $\mathbf{p}_i$ measured with respect to a reference system located in the center of mass (CM) of the object. Each point has also an associated local coordinate system, where $\mathbf{n}_i$ is the unitary inward pointing normal, and $\mathbf{s}_i$ and $\mathbf{t}_i$ are the tangential vectors such that $\mathbf{n}_i = \mathbf{s}_i \times \mathbf{t}_i$ is satisfied (Figure 3.1a). N is assumed to be large enough to accurately represent the object.

A grasp is defined as a set $\mathcal{C}$ of n contact points and a set $\mathcal{F}$ of n corresponding contact forces that are applied by the robotic hand on the object surface. The contact between each finger and the object can be modeled as a punctual contact (with or without friction) or as a soft finger contact. In general, the contact force $\mathbf{f}_i$ is expressed in the coordinate system of the contact point as $\mathbf{f}_i = \begin{bmatrix} f_{i,1} & f_{i,2} & f_{i,3} & f_{i,4} \end{bmatrix}^T$, where $f_{i,1}$ is the normal component along $\mathbf{n}_i$, $f_{i,2}$ and $f_{i,3}$ are the tangential components along $\mathbf{s}_i$ and $\mathbf{t}_i$ respectively, and $f_{i,4}$ is the moment around $\mathbf{n}_i$. To ensure a stable contact without slippage, the contact force needs to lie inside the friction cone, $\mathbf{f}_i \in F_i$, which describes the positivity

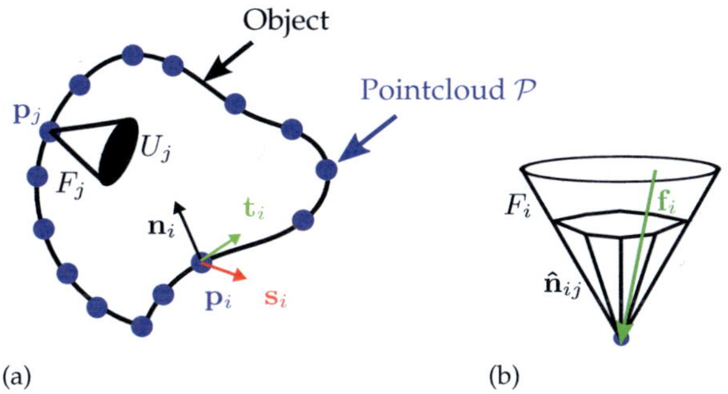

Figure 3.1.: (a) The object surface is represented by a pointcloud, and each point has an associated local coordinate system where $\mathbf{n}_i$ is the inward pointing normal. The friction cone F_j can either be described as the convex cone of the set U_j or (b) by linearizing it with an m-side polyhedral convex cone. Note that in both cases the magnitude of the normal componant is normalized to one.

and friction constraints for the chosen contact model [144]. For instance, for the punctual contact with friction (PCwF), the friction cone is described as

$$F_i = \left\{ f_{i,1} \geq 0, \sqrt{f_{i,2}^2 + f_{i,3}^2} \leq \mu f_{i,1}, f_{i,4} = 0 \right\}, \tag{3.1}$$

where μ is the friction coefficient. Furthermore, the friction cone can be described as the convex cone of a set U_i, which depends on the chosen contact model [234]. For the PCwF model, the set U_i is

$$U_i = \left\{ \mathbf{f}_i \mid f_{i,1} = 1, \sqrt{f_{i,2}^2 + f_{i,3}^2} = \mu, f_{i,4} = 0 \right\}. \tag{3.2}$$

Note that the set uses $f_{i,1}$ normalized to one [234]. In the remainder of the section, the PCwF model is employed, although the proposed algorithms can also be used with frictionless point contacts or soft finger contacts. In the 3-dimensional space, this model is nonlinear and, to simplify it for the computation of the independent contact regions (Section 3.3), the friction cone is linearized using an m-side polyhedral convex cone. Thus, by representing the unitary vector along the j-th edge of the convex cone at the i-th contact with $\hat{\mathbf{n}}_{ij}$, the grasping force is

$$\mathbf{f}_i = \sum_{j=1}^{m} \alpha_{ij} \hat{\mathbf{n}}_{ij} , \quad \alpha_{ij} \geq 0, \quad \|\hat{\mathbf{n}}_{ij}\| = 1. \tag{3.3}$$

The force $\mathbf{f}_i$ applied on the object at $\mathbf{p}_i$ generates a torque $\boldsymbol{\tau}_i = \mathbf{p}_i \times \mathbf{f}_i$ with respect to CM. $\mathbf{f}_i$ and $\boldsymbol{\tau}_i$ are grouped together in a wrench vector given by

$\boldsymbol{\omega}_i = \left(\mathbf{f}_i^T\ \boldsymbol{\tau}_i^T\right)^T$. The wrench $\mathbf{w}_{ij} = \left(\hat{\mathbf{n}}_{ij}^T,\ (\mathbf{p}_i \times \hat{\mathbf{n}}_{ij})^T\right)^T$ generated by a unitary force $\hat{\mathbf{n}}_{ij}$ along the edge j of the linearized friction cone is called a *primitive wrench* (Figure 3.1b). The set W_i contains the primitive contact wrenches of the linearized friction cone $W_i = \{\mathbf{w}_{i1}, \ldots, \mathbf{w}_{im}\}$.

The wrench $\boldsymbol{\omega}_i$ can also be computed as $\boldsymbol{\omega}_i = \mathbf{G}_i \cdot \mathbf{f}_i$. Here, $\mathbf{G}_i \in \mathbb{R}^{6\times 4}$ gives the transformation from the coordinate system of the contact point to the object coordinate system and is defined as

$$\mathbf{G}_i = \begin{bmatrix} \mathbf{n}_i & \mathbf{s}_i & \mathbf{t}_i & \mathbf{0} \\ \mathbf{p}_i \times \mathbf{n}_i & \mathbf{p}_i \times \mathbf{s}_i & \mathbf{p}_i \times \mathbf{t}_i & \mathbf{n}_i \end{bmatrix}. \tag{3.4}$$

The matrix $\mathbf{G}_i$ can also be used to obtain the set W_i of primitive wrenches that can be applied at a contact point without linearizing the friction cone: $W_i = \mathbf{G}_i(U_i)$ [234]. The overall wrench $\boldsymbol{\omega}$ acting on the object can then be expressed by:

$$\boldsymbol{\omega} = \mathbf{G} \cdot \mathbf{f} = \begin{bmatrix} \mathbf{G}_1 & \mathbf{G}_2 & \ldots & \mathbf{G}_n \end{bmatrix} \cdot \begin{bmatrix} \mathbf{f}_1 \\ \mathbf{f}_2 \\ \vdots \\ \mathbf{f}_n \end{bmatrix}, \tag{3.5}$$

with $\mathbf{f} \in \mathbb{R}^{4n}$, and $\mathbf{G} \in \mathbb{R}^{6\times 4n}$ being the grasp matrix.

A force closure (FC) grasp must be able to counteract any external wrench $\boldsymbol{\omega}_{\text{ext}}$ that acts on the object, i.e. $\boldsymbol{\omega} = -\boldsymbol{\omega}_{\text{ext}}$ [144]. In other words, any $\boldsymbol{\omega}_{\text{ext}}$ can be expressed as a positive combination of contact wrenches or, geometrically speaking, any $\boldsymbol{\omega}_{\text{ext}}$ lies within the set of all possible wrenches acting on the object. A grasp is then said to be force closure if and only if the origin of the wrench space lies strictly inside the set of all possible wrenches, computed as the convex hull $CH(W)$ of a set W that contains the primitive wrenches for the current grasp [61, 144]. This implies that a ray emerging from the origin along the direction of $\boldsymbol{\omega}_{\text{ext}}$ has always one (and only one) intersection point with $CH(W)$. Hence, the FC test is not dependent on the magnitudes of the external wrench or the contact forces, but can be checked using the sets of primitive contact wrenches W_i. Under the assumption that the magnitude of the overall contact force is bounded, i.e. the robotic hand has one limited power source, the set W can be derived as the union of all convex cones W_i

$$W = W_{L_1} = W_1 \cup \ldots \cup W_n. \tag{3.6}$$

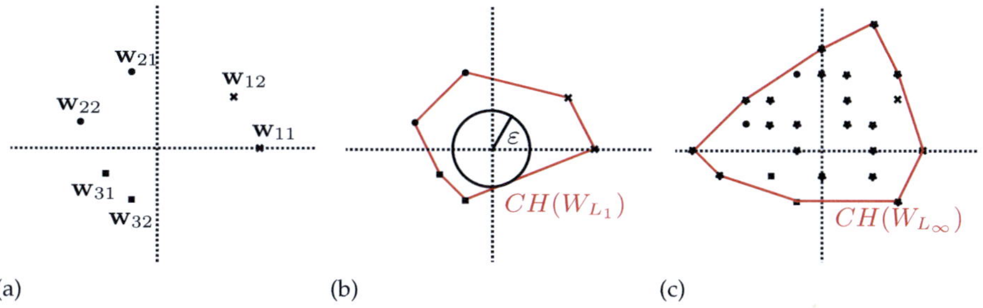

Figure 3.2.: (a) An abstract 2-dimensional wrench space for a three-finger frictional grasp. Two primitive wrenches $W_i = \{\mathbf{w}_{i1}, \mathbf{w}_{i2}\}$ are defined per contact point i. (b) The convex hull of the set $W_{L_1} = \cup_i W_i$ (c) The convex hull of the set $W_{L_\infty} = \oplus_i W_i$.

Contradictory, physical hands usually have independent power sources for every finger, i.e. the magnitude of each contact force is independently bounded. In this case, W is defined as

$$W = W_{L_\infty} = W_1 \oplus \ldots \oplus W_n, \tag{3.7}$$

with $\oplus$ being the Minkowski sum. For two arbitrary subsets $W_1, W_2 \subset \mathbb{R}^m$, the Minkowksi sum is [155]

$$W_1 \oplus W_2 = \{\mathbf{c} = \mathbf{a} + \mathbf{b} \,|\, \mathbf{a} \in W_1, \mathbf{b} \in W_2\}. \tag{3.8}$$

Figure 3.2 shows an abstract 2-dimensional wrench space, the corresponding sets W_{L_1} and W_{L_∞}, and their convex hulls. Although W_{L_∞} uses more realistic assumptions than W_{L_1}, typically W_{L_1} is used to verify the FC property as it is simpler to compute. The result is a conservative estimate of the FC property. Also the approach used for the computation of independent contact regions relies on W_{L_1} and a linearized friction cone (see Section 3.3). However, for the computation of contact forces in Section 3.6, we overcome the linearization of the friction cone, the assumption of normalized contact forces, and use W_{L_∞} for a realistic estimation of the capacities of the robotic hand.

Besides the requirement of force closure, which leads to a stable grasp, grasps can analytically be evaluated with a certain quality [203]. Mostly the ability of the grasp to counteract an external disturbance wrench is rated. Usually the direction of the external wrench $\boldsymbol{\omega}_{\text{ext}}$ is not known a-priori and hence, all directions are taken into account by measuring the radius ε of the largest inscribed ball in $CH(W)$ as shown in Figure 3.2b. Note that this widely used ε-metric is purely based on geometrical considerations of contact points and normalized contact forces.

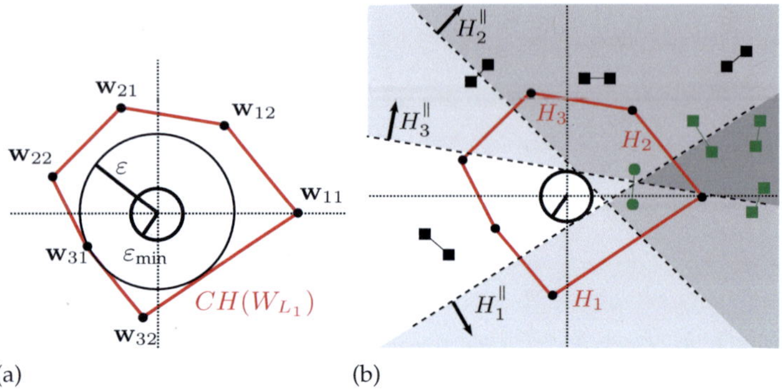

Figure 3.3.: (a) An abstract 2-dimensional wrench space for a three finger frictional grasp. Two primitive wrenches $W_i = \{\mathbf{w}_{i1}, \mathbf{w}_{i2}\}$ are defined per contact point i and the convex hull of the set $W_{L_1} = \cup_i W_{\text{init},i}$ is shown in red. (b) The dotted lines show the shifted hyperplanes for the initial contact point corresponding to finger 1 and the grey regions correspond to their spanned half-plane. The wrenches depicted in green belong to $\mathcal{ICR}_1$.

3.3. Independent Contact Regions

The basic approach used in this thesis allows for a fast computation of the independent contact regions (ICRs) and is a reformulation of the algorithms presented in [46, 172]. The surface of the object is described with a set $\mathcal{P}$ of points and a friction coefficient μ that estimates the friction between the fingertips and the object. The algorithm uses geometrical relations in the wrench space starting from an initial force closure grasp $\mathcal{C}_{\text{init}} = \{\mathbf{g}_1, \ldots, \mathbf{g}_n\}$, $\mathbf{g}_i \in \mathcal{P}$. The corresponding set of primitive wrenches $W_{\text{init},i}$ per finger i is obtained from linearized friction cones (with m sides) at the contact points. Figure 3.3a shows an abstract 2-dimensional wrench space. The black circles are the corresponding primitive wrenches for an initial three finger frictional grasp where the friction cone is represented by two primitve wrenches per contact point. The set of all possible wrenches acting on the object is represented by the convex hull of $W_{L_1} = W_{\text{init}} = \cup_i W_{\text{init},i}$ (the red line in Figure 3.3a). The boundary of the convex hull can also be described by intersecting hyperplanes H_j that are spanned by the primitive wrenches that lie on the boundary. These supporting hyperplanes are used to calculate the ε quality of the grasp by computing the distance of each hyperplane to the origin of the wrench space. The smallest distance determines the ε quality. In a next step, the supporting hyperplanes are sorted into sets $\mathcal{K}_i$.

Algorithm 1: Computation of ICRs

Given:

- set of points $\mathcal{P}$ describing the object surface and corresponding wrenches W_i from the linearized friction cones
- friction coefficient μ
- ratio α for a minimum grasp quality $\varepsilon_{\text{min}} = \alpha\varepsilon$ for all grasps contained in the ICRs
- initial force closure grasp $C_{\text{init}} = \{\mathbf{g}_1, \ldots, \mathbf{g}_n\}$, $\mathbf{g}_i \in \mathcal{P}$

Output: $\mathcal{ICR} = \{\mathcal{ICR}_1, \ldots, \mathcal{ICR}_n\}$

$W_{\text{init}} \leftarrow \{W_{\text{init},1} \cup \ldots \cup W_{\text{init},n}\}$;
Obtain $CH(W_{\text{init}})$ and grasp quality ε;
foreach *contact point* $\mathbf{g}_i$ **do**
    $\mathcal{K}_i \leftarrow$All the supporting hyperplanes of $CH(W_{\text{init}})$ that contain at least one primitive wrench of $\mathbf{g}_i$;
    $\mathcal{K}_i^{\parallel} \leftarrow$All hyperplanes parallel to $\mathcal{K}_i$ and tangent to the sphere of radius ε_{min};
    Create a queue $\mathcal{Q}$;
    Enqueue $\mathbf{g}_i$ into $\mathcal{Q}$;
    while *$\mathcal{Q}$ is not empty* **do**
        $\mathbf{p}_t \leftarrow \mathcal{Q}.dequeue()$;
        if *inclusionTest*($\mathbf{p}_t$, $\mathcal{K}_i^{\parallel}$); `// Algorithm 2`
        then
            $\mathcal{ICR}_i \leftarrow \mathbf{p}_t$;
            forall the *points* $\mathbf{p}_v$ *neighbor of* $\mathbf{p}_t$ **do**
                if $\mathbf{p}_v$ *is not marked* **then**
                    Mark $\mathbf{p}_v$;
                    Enqueue $\mathbf{p}_v$ into $\mathcal{Q}$;

A hyperplane H_j belongs to $\mathcal{K}_i$ if it contains at least one primitive wrench of the i-th contact point $\mathbf{g}_i$ (H_1, H_2, and H_3 in Figure 3.3b). After sorting, all hyperplanes of the set $\mathcal{K}_i$ are shifted such that they are tangent to the sphere of radius $\varepsilon_{\text{min}} = \alpha\varepsilon$, where $0 < \alpha \leq 1$ is a given factor defined beforehand that specifies a minimum desired ε-quality for the ICRs. The shifted hyperplanes $H_k^{\parallel}$ are stored in the set $\mathcal{K}_i^{\parallel}$ and the surface point $\mathbf{g}_i$ is the first point that belongs to the independent contact region $\mathcal{ICR}_i$. The region is expanded in a next step by exploring the neighboring surface points of $\mathbf{g}_i$ with a breadth-first search. A point belongs to the contact region if it is ensured that the force closure property with the contact points of the other fingers is fulfilled and the grasp has a minimum grasp quality of ε_{min}. This is tested with the positive halfspace of the shifted hyperplanes $H_k^{\|}$ in $\mathcal{K}_i^{\|}$ that points away from the origin of the wrench space. A point is included in $\mathcal{ICR}_i$ if there is at least one corresponding primitive wrench in each $H_k^{\|+}$ in $\mathcal{K}_i^{\|}$. Figure 3.3b shows the positive halfspaces in grey and primitive wrenches of points that belong to $\mathcal{ICR}_i$ in green. Green squares represent contact points where at least one primitive wrench lies in the intersecting region $H_1^{\|+} \cap H_2^{\|+} \cap H_3^{\|+}$, while green circles depict a point whose primitive wrenches do not lie in the intersection region but still fulfill the condition of having at least one primitive wrench inside each positive half-space, therefore belonging to the $\mathcal{ICR}_i$. Algorithm 1 summarizes the concept of the independent contact regions and Algorithm 2 details the inclusion test.

Algorithm 2: Inclusion test for ICRs

Given:

- tested point $\mathbf{p}_t$
- shifted hyperplanes $\mathcal{K}_i^{\|}$ that contain at least one primitive wrench of $\mathbf{g}_i$

Output: a boolean (true if $\forall H_k^{\|} \in \mathcal{K}_i^{\|}, \exists \omega_{tj} \in H_k^{\|+}, j = 1, \ldots, m$)

foreach $H_k^{\|} \in \mathcal{K}_i^{\|}$ **do**
    contain ← false;
    $H_k^{\|+}$ ← positive side of $H_k^{\|}$;
    forall the *primitive wrench* ω_{tj} *of* $\mathbf{p}_t$ **do**
        if $\omega_{tj} \subset H_k^{\|+}$ **then**
            contain ← true;
            break ; `/* go to next plane */`
    if *contain is false* **then**
        return false;
return true;

3.4. The Workspace of the Robotic Hand

3.4.1. Concept

The workspace of the robotic hand depends only on its kinematic configuration and can therefore be computed offline as it does not change during the planning process. The workspace presented in this section is later used for the online computation of reachable points on the object (Section 3.5), and also to quickly obtain the solution of the inverse kinematics, and hence, the forces that the finger can apply on the surface of the object (Section 3.6).

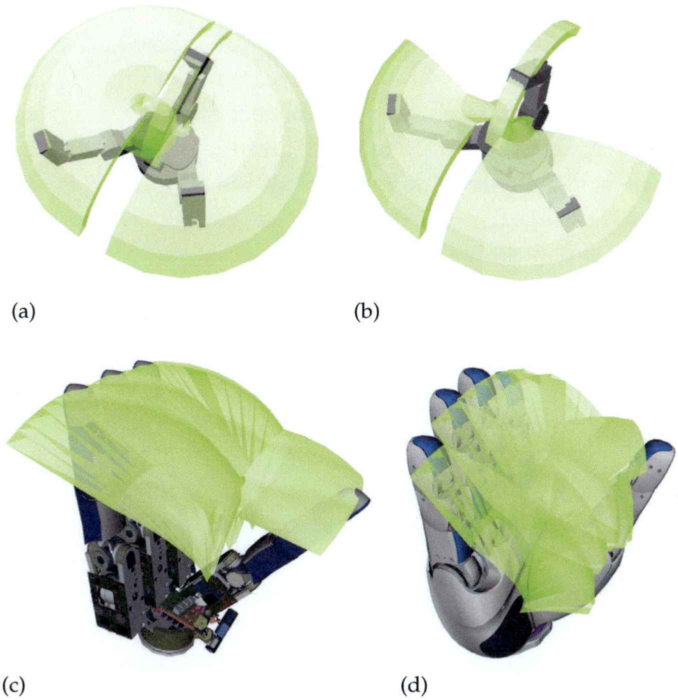

Figure 3.4.: Workspaces for several robotic hands (a) Set Φ_{ft} of reachable points for the fingertips of the Barrett Hand (b) The corresponding set Φ_{fc} of points that potentially lead to FC grasps for the Barrett Hand. (c) Set Φ_{ft} for the DLR Hand (d) Set Φ_{ft} for the DLR-HIT Hand II.

A multi-fingered hand can be considered as a collection of small manipulators (fingers) with a common base (palm), and the hand workspace is thus the union of the workspaces of each finger. In other words, the workspace Φ of the robotic hand is the set of all spatial points, located with respect to the base coordinate system of the hand, that are reachable for at least one possible hand configuration. For a hand with n fingers, the workspace is $\Phi = \phi_1 \cup \phi_2 \cup \ldots \cup \phi_n$, with ϕ_i the workspace of the i-th finger. Note that there is no restriction on the part of the hand that reaches a point in Φ, i.e. potential contact points with any part of the finger are possible.

This work is centered on precision grasps, therefore only the points reachable with the tips of the fingers must be considered, which defines the set $\Phi_{ft} \subset \Phi$. We are also interested in obtaining fingertip FC grasps, which leads to the definition of a new set of points $\Phi_{fc} \subset \Phi_{ft}$ reachable for the fingertips and that potentially allow an FC grasp; $\Phi_{fc} = \phi_{fc,1} \cup \ldots \cup \phi_{fc,n}$. In practice, $\phi_{fc,i}$ can be obtained from $\phi_{ft,i}$ by trimming the parts of the fingertip workspace resulting from finger positions not useful for precision grasps. For this, the hand configuration space is uniformly sampled, and the position of the fingertip region is computed via the direct kinematics of the hand. The FC condition can be tested using the normals to the fingertips, a predefined coefficient of friction μ_p, and a common coordinate system with its origin located at the centroid of the considered fingertip points. The artificially imposed μ_p should be the maximal value expected for the applications of the real hand-object system. As an example, Figure 3.4a shows for a Barrett hand [207] the set of reachable points Φ_{ft} for a patch defined on each fingertip. Figure 3.4b shows for the same hand the set Φ_{fc}, computed with a friction coefficient of $\mu_p = 0.5$.

Additional restrictions can be considered to further restrain the workspace of the hand considered in the online stage. In particular, avoiding self-collisions is a desirable property, although the workspace that guarantees no self-collisions for the case of anthropomorphic hands would be so reduced that it renders this consideration impractical [251]. For this work, the set Φ_{fc} is taken as the useful portion of the hand workspace in case of hands with non-modular fingers. In case of an anthropomorphic hand like the DLR Hand II [38] or the DLR-HIT Hand II [124], Φ_{ft} is used to restrict the capabilities of the hand as little as possible (Figure 3.4c and 3.4d). This set is used for a fast check of reachability for points on the object surface.

In [251], we presented a workspace representation of the robotic hand, which provided all the points that each fingertip could reach for one predefined point on its surface. The assumption of one contact point per fingertip was an over simplification of the capabilities of the hand. In this work, we assume that the fingers are not kinematically redundant, and the fingertip is approximated as a sphere. The end of the kinematic chain is then the central point of the fingertip sphere as shown in Figure 3.5.

To consider not only reachability of the grasp but also feasibility, the achievable contact forces of the robotic hand must be taken into account. Those forces depend on the joint configuration of the finger, i.e. they change within the hand workspace, and additionally, on the direction of the normal at the desired contact point. Figure 3.6b depicts two exemplary contact points and their maximal applicable forces (in yellow) for one joint configuration. Besides, these forces need to be quickly obtained to guarantee that the grasp planning can be

(a) (b) (c)

Figure 3.5.: The upper part of the fingertip of the DLR-HIT Hand showing the approximation with a sphere (in green). The parts of the fingertip considered for grasping are shown in black.

executed with a low computation time. This is realized by storing the solution to the inverse kinematics for the finger, as detailed in the next paragraph.

3.4.2. Implementation Details

For a robotic manipulator, the representation of the regions where the Tool Center Point (TCP) can be moved to is known as a reachability map [226]. It can be computed as a grid in the 6D space (position and orientation) where each cell has a binary value that indicates if it is reachable or not. The same approach is used here to compute the reachable workspace for the fingers. The reachability map for each finger is computed offline using a hybrid approach that combines forward and inverse kinematics to obtain an accurate and structured description of the finger capabilities [166]. The finger configuration leading to each point in space is also computed and stored for later use in the verification of directional reachability for the points on the object and to easily obtain the finger pose leading to some particular goal contact point. With a high resolution of the sampled grid, the finger joint configurations that reach the cartesian space enclosed within one grid step are not significantly different.

The workspace of each finger of the DLR-HIT Hand II [124] is voxelized with 0.001 m voxel size which results in a bounding box including 573400 voxels, where 438438 voxels are not reachable and 134962 voxels are reachable. For the reachable voxels, the corresponding joint configurations are stored. Using [166], the offline computation of the workspace takes about 6 min and has to be calculated only once. The computation of the workspace can also be done by sampling the possible joint configurations and storing each reachable voxel. It

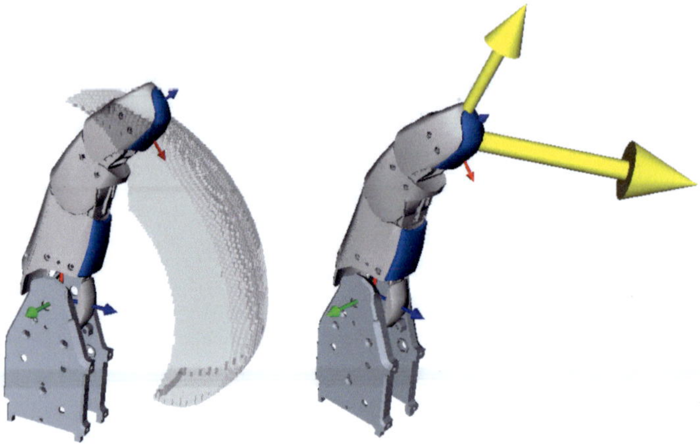

(a) (b)

Figure 3.6.: (a) Workspace of one finger of the DLR-HIT Hand II, shown with a particular joint angle configuration. The workspace describes the reachable positions for the end of the forward kinematics which is marked by the upper coordinate system. (b) The magnitudes of forces that the finger can apply, even for the same joint configuration, may change depending on the particular contact point and the direction of the required force. Two exemplary contact points and their maximal applicable forces are shown here in yellow.

leads to the same result as using [166] but is computationally more expensive. The reachable voxels are shown in grey for the thumb in Figure 3.6a. In the shown workspace, the end of the forward kinematics is in the middle of the sphere that represents the fingertip. The radius of the sphere is 0.011 m. The approximation of the fingertip as a sphere is only valid for parts of the fingertip as shown in Figure 3.5. We restrict the reachability to the black parts of the fingertip.

This representation of the workspace is later used during the planning process in order to quickly obtain the intersection between object and workspace, i.e. the reachable points on the object surface, and the joint configurations of the fingers that lead to those reachable points. This intersection is computed with an enhanced version of the Voxmap-Pointshell (VPS) algorithm [181, 182]. The VPS is a haptic rendering algorithm providing fast responses to collision queries even for arbitrarily complex scenarios. This is realized by using two datastructures: *voxmaps* and *pointshells* as depicted in Figure 3.7a. Voxelmaps are 3D grids in which each voxel stores a discrete distance value $s \in \mathbb{Z}$ to the surface. Voxels on the surface layer have $s = 0$, voxels in the kth inner layer

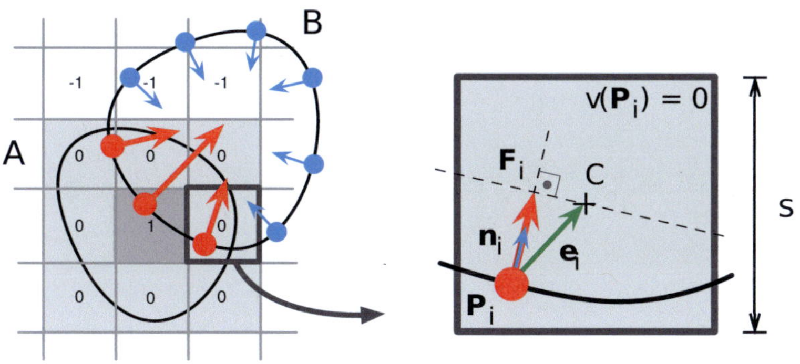

(a) (b)

Figure 3.7.: (a) Datastructures of the Voxmap-Pointshell (VPS) algorithm. Object A is a voxmap where the surface is represented as voxels with layer 0. Object B is represented as pointshell depicted in blue. (b) Computation of the collision force generated by a single colliding point.

$s = k$, and voxels in the kth outer layer have $s = -k$. Additionally, the scalar voxelmap function $V(\mathbf{p})$ yields the signed distance value of a point $\mathbf{p}$ in the voxelmap. Pointshells are sets of points uniformly distributed on the surface of the object; each point $\mathbf{p}_i$ has additionally an inwards pointing normal vector $\mathbf{n}_i$. The total collision force and torque ($\mathbf{f}_{\text{tot}}$, $\mathbf{t}_{\text{tot}}$) for each colliding object-pair is computed as the sum of all collision forces and torques ($\mathbf{f}_i$, $\mathbf{t}_i$) generated by colliding points $\mathbf{p}_i$ (Figure 3.7b).

Points are colliding if $s \geq 0$, and their normal vectors $\mathbf{n}_i$ are weighted by their penetration in the voxelmap $V(\mathbf{p}_i)$, resulting in the collision force $\mathbf{f}_i$. Torques $\mathbf{t}_i$ generated by colliding points are the cross product between forces $\mathbf{f}_i$ and point coordinates $\mathbf{p}_i$, all magnitudes expressed in the pointshell frame, with its origin in the center of mass. A detailed description of the use of the VPS during the grasp planning is given in the next sections.

3.5. Reachable Independent Contact Regions

The algorithm of Roa et al. [172], presented in Section 3.3, is a search for contact regions independent of a robotic hand. Hence, the contact points are computed in order to fulfill force closure but they might not be reachable by a real robotic hand. This section considers the problem of computing contact regions

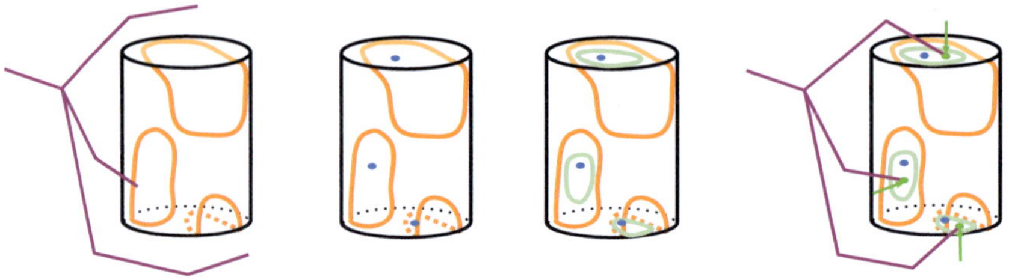

Figure 3.8.: Process to find rICRs on the object: (a) Reachable regions for the fingertips on the object surface (b) Initial FC grasp on the object (c) rICRs on the object (d) Goal contact points and a corresponding hand configuration to provide a robust FC grasp.

that can be reached by a robotic hand for the application in a shared autonomy setup. Computing optimal contact regions requires more computational power and has been tackled in [46, 252]. Figure 3.8 schematically describes the complete approach. First, reachable points on the object surface are computed using a given hand (palm) pose relative to the object and a representation of the hand workspace. Then, a first force closure (FC) grasp on the object is obtained among the reachable points. This initial FC grasp serves as the basis to expand the reachable independent contact regions (rICRs) on the object surface. Finally, goal contact points inside the rICRs are chosen, to decide the best finger configuration for obtaining a robust grasp on the object.

This computation of rICRs is realized in an offline and an online stage. The offline stage was presented in the previous section and deals with the computation of a suitable hand workspace Φ that describes the configurations of the robotic hand that will potentially lead to good grasps. The online stage looks for the intersection between the workspace Φ and the set $\mathcal{P}$ of points describing the object boundary, and uses this information for the computation of rICRs on the object.

3.5.1. Algorithm

The online computation of rICRs requires, like the computation of IRCs, a set $\mathcal{P}$ of points describing the object surface, and a friction coefficient μ that estimates the friction between the fingertips and the object. Additionally, a hand pose (position and orientation) relative to the object is used as an input. The search for an initial FC grasp is described as well in this section.

The basis for the rICRs are the reachable points for each fingertip on the object. They are computed as the intersection ψ_i between the useful workspace of the finger and the point cloud of the object, i.e. $\psi_i = \phi_{fc,i} \cap \mathcal{P}$. A detailed description of the positional and directional reachability verification is provided in Section 3.5.2. The point $\mathbf{g}_i \in \psi_i$ closest to each fingertip is computed and these points are used to form an initial grasp $\mathcal{C}_{\text{init}}$. The FC property on this initial guess is tested via a simplified ray-shooting test [232] which provides a fast answer to the query. If $\mathcal{C}_{\text{init}}$ is not FC, a search is performed for one contact point at a time to try and relocate it inside its corresponding set ψ_i to get an FC grasp with the remaining $n-1$ contact points, following the algorithm proposed in [172]. Algorithm 3 summarizes the procedure.

If $\psi = \{\psi_1, \ldots, \psi_n\}$ is empty (i.e. there are no reachable points for the current hand configuration), if there is only one finger with a non-empty ψ_i, or if no initial FC grasp is found, then the relative hand pose will not provide rICRs.

Algorithm 3: Search for an initial FC grasp

Given:

- set of points $\mathcal{P}$ describing the object surface
- suitable subset Φ of the hand workspace

Output: $\psi = \{\psi_1, \ldots, \psi_n\}$, $\mathcal{C}_{\text{init}}$

foreach *Finger i* **do**
  Compute $\psi_i \leftarrow \phi_i \cap \mathcal{P}$;
  if $\psi_i = \emptyset$ **then**
    Compute closest point $\mathbf{p}_i$ in $\mathcal{P}$; `/*` $\mathbf{p}_i$ `is not reachable and is not considered for` $\mathcal{C}_{\text{init}}$ `*/`
  else
    Compute closest point $\mathbf{g}_i$ in ψ_i;

Form the initial grasp $\mathcal{C}_{\text{init}} = \{\mathbf{g}_1, \ldots, \mathbf{g}_l\}$, and obtain the set $\mathbf{W}_{\text{init}}$ of primitive wrenches ; `/*` $\mathcal{C}_{\text{init}}$ `must contain at least` $l = 2$ `reachable points */`

Check force closure ; `/* Uses a ray-shooting test, [232] */`

while *$\mathcal{C}_{init}$ is not FC* **do**
  look for a new grasp $\mathcal{C}_{\text{init}}$, changing a contact point $\mathbf{g}_i \in \psi_i$ at a time ; `/* Uses algorithm from [172] */`

In some particular applications, like using rICRs as a visual help in telemanipulation for grasping known objects (as in Section 5.4, [243]) the closest, but not reachable, points $\mathbf{p}_i$ to each fingertip (Line 4 in Algorithm 3) are then used as an initial guess to find the closest FC grasp (following again the algorithm in [172]), to provide it as a suggestion for the user so that the hand posture can be adapted to try to reach those points.

If an initial FC grasp $\mathcal{C}_{\text{init}}$ is found with Algorithm 3, the contact regions are computed using Algorithm 4. To guarantee reachable ICRs, the breadth-first search in the neighboring surface points of $\mathcal{C}_{\text{init}}$ (Line 13 in Algorithm 1) is altered. Every new point $\mathbf{p}_v$ potentially belonging to $\mathcal{ICR}_i$ needs to be among the set of reachable points ψ_i.

Algorithm 4: Computation of reachable ICRs and a goal grasp $\mathcal{C}_{\text{goal}}$

Given:

- set of points $\mathcal{P}$ describing the object surface and corresponding wrenches W_i from the linearized friction cones
- friction coefficient μ
- ratio α for a minimum grasp quality $\varepsilon_{\min} = \alpha\varepsilon$ for all grasps contained in the ICRs

Output: $r\mathcal{ICR} = \{r\mathcal{ICR}_1, \ldots, r\mathcal{ICR}_n\}$, $\mathcal{C}_{\text{goal}} = \{\mathbf{p}_1, \ldots, \mathbf{p}_n\}$

Obtain reachable points ψ and an initial force closure grasp $\mathcal{C}_{\text{init}} = \{\mathbf{g}_1, \ldots, \mathbf{g}_n\}$; `// Algorithm 3`
$W_{\text{init}} \leftarrow \{W_{\text{init},1} \cup \ldots \cup W_{\text{init},n}\}$;
Obtain $CH(W_{\text{init}})$ and grasp quality ε;
foreach *contact point* $\mathbf{g}_i$ **do**
    Get $\mathcal{K}$ and $\mathcal{K}_i^{\parallel}$;
    Create a queue $\mathcal{Q}$ and enqueue $\mathbf{g}_i$;
    while *$\mathcal{Q}$ is not empty* **do**
        $\mathbf{p}_t \leftarrow \mathcal{Q}.dequeue()$;
        if *inclusionTest*($\mathbf{p}_t$, $\mathcal{K}_i^{\parallel}$); `// Algorithm 2`
        then
            $r\mathcal{ICR}_i \leftarrow \mathbf{p}_t$;
            forall the *points* $\mathbf{p}_v$ *neighbor of* $\mathbf{p}_t$ **do**
                if *($\mathbf{p}_v$ is not marked) and ($\mathbf{p}_v \in \psi_i$)* **then**
                    Mark $\mathbf{p}_v$;
                    Enqueue $\mathbf{p}_v$ into $\mathcal{Q}$;
Compute the goal contact points $\mathcal{C}_{\text{goal}}$

As a last step for grasping, goal contact points C_{goal} need to be chosen within the computed rICRs. Different criteria can be used here, for instance, selecting the points that provide maximum grasp quality, the centroid of each rICR, or contact points that lead to finger joint configurations as far from singularities as possible. A primary goal is to avoid collisions between the fingers while reaching the goal contact points. Hence, a collision free path from initial finger joint angles to a final configuration somewhere in the rICRs would be ideally planned. Instead of this computationally expensive step, two checks are performed to limit the probability of collision. First, potential intersections between the computed rICRs are verified. If there is some overlapping between a pair of rICRs, the points lying in the intersection are divided between the two conflicting rICRs to create disjoint sets. Then, the centroid of each disjoint rICR is chosen as goal contact point, to increase the likeliness of the real contact points being within the computed contact region despite positioning errors (due, for instance, to errors in the forward kinematics of the hand, or on the object location with respect to the hand). The finger joint angles are obtained to place the fingers at these contact points (Section 3.4) and a collision check between the fingertips is performed. If the fingers collide, the goal contact points are moved away from each other to avoid collision.

Finally, note that there is no explicit mention of a particular contact model (frictionless grasp, punctual frictional grasp or soft finger contacts [144]), which influences how the primitive wrenches in W are computed. As the rICR approach works directly on the wrench space (i.e. it only considers the set W of primitive contact wrenches for the computation), it is general enough to be used with any contact model. The overall algorithm is summarized in Algorithm 4.

Complexity

The computation of the initial FC grasp uses a simplified ray-shooting approach [232] which provides a fast boolean answer by solving an LP problem. The computation of the convex hull $CH(W_{\text{init}})$ in Step 2 of Algorithm 4 uses the Qhull-package [12]. For a 6-dimensional input, it has a complexity of $\mathcal{O}((n_v \cdot m)^3/6)$ where n_v is the number of contact points whose associated m primitive wrenches form the vertices of $CH(W_{\text{init}})$. Regarding the breadth-first-search required to build the rICRs, since each neighbor for each contact i is inspected, the complexity is $\mathcal{O}(N_{r,i} \cdot n)$, with $N_{r,i}$ the total number of points inside the corresponding reachable region ψ_i. Hence, the overall complexity is $O(((n_v \cdot m)^3/6) + N_r \cdot n)$.

3.5.2. Implementation Details

One of the critical parts for the implementation of rICRs is precisely the computation of the sets ψ of reachable points on the object surface, resulting from the intersection of the finger workspaces with the object (Step 1 in Algorithm 4). In the rICRs implementation, the pointshell for the VPS algorithm corresponds to the set $\mathcal{P}$ of points on the object boundary (Figure 3.9a). The voxmap is the finger workspace, computed as described in Section 3.4.2. The outcome of the VPS algorithm includes the sets ψ_i of intersecting points on the object, and the corresponding colliding voxels.

As the voxmap index indicates the finger joint configuration that leads to the middle of the fingertip being on the object surface, the correct voxel is found by moving backwards, following the negative direction of the surface normal, with a distance equal to the radius of the fingertip (Figure 3.9b). If the new point results in a voxel that is outside of the finger workspace, the point on the object surface is discarded.

This initial result accounts for positional reachability, i.e. the intersecting points are within the workspace of the finger. However, an additional test must be

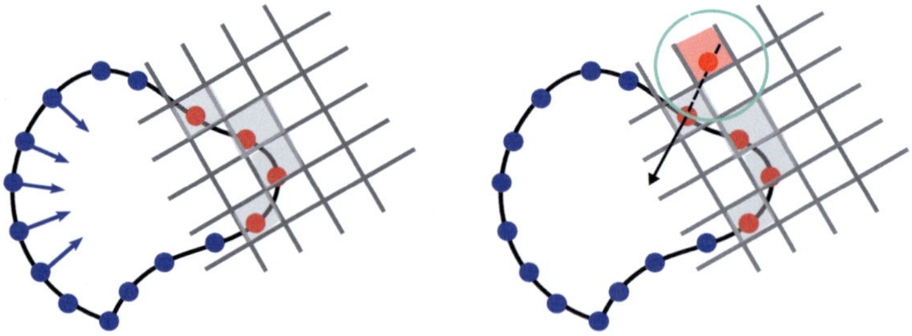

(a) (b)

Figure 3.9.: (a) The object is represented as a pointshell (blue points and surface normals) and the workspace of the finger is represented as a voxmap (grey grid). The colliding pointshell points are depicted in red, and the shaded boxes show the colliding voxels on the surface of the object (b) Obtaining the correct voxel (shaded red) that leads to the fingertip being on the object surface. In cyan: the fingertip represented as a sphere.

performed to verify the directional reachability, which guarantees that the fingertip can apply forces within the friction cone at the possible contact points. To verify this condition, the fingertip normal at each potential contact point $\mathbf{p}_i$ is computed using the stored finger configuration at the corresponding voxel. Then, the angle θ between the fingertip normal and the corresponding surface normal at the contact point $\mathbf{p}_i$ is computed, and the point is also considered reachable in direction if $\theta \leq 2\,\mathrm{atan}(\mu)$. An area of possible contact points on the spherical fingertip can also be considered for this reachability verification.

The VPS algorithm is additionally used to compute the closest point on the object with respect to each fingertip (Line 4 in Algorithm 3) and to check for the collision of the fingertips in the goal configuration of the hand. For this purpose, the last link of each finger is voxelized offline. The voxmap size is artificially increased by expanding the outer layer of voxels, i.e. creating a distance field around the fingertips. Checking the collision of the pointshell of the object with these layers allows a fast detection of the closest point on the object.

3.5.3. Evaluation

The proposed approach was implemented using Matlab/Simulink and C++. Computational times were evaluated on a standard Linux computer with an Intel® Xeon® CPU E5-1620 v2 processor running at 3.7 Ghz. The calculation of rICRs is analyzed for the two most time-consuming parts: computation of reachable points on the object surface, and calculation of the contact regions.

The computation of reachable points is executed in parallel using a separate thread for each finger. Therefore, this calculation time mainly depends on the number N of points in the set Ω. The influence of the computation of reachable points is analyzed by choosing a fixed pose for the DLR-HIT Hand II [124] relative to an industrial object, as illustrated in Figure 3.10a, where the workspace of the thumb in shown in green. Different resolutions for discretization of the object are used, with $N \in \{3104, 4564, 6606, 10733, 18350\}$, which correspond to a point every 3.5, 2, 2.5, 2, and 1.5 mm on the object surface, respectively. Additionally, the size of the region on the fingertip that is allowed to contact the object is changed, to provide different numbers of reachable points. Figure 3.11 shows the computational time required to find reachable points on the object surface under these conditions; the figure shows the result of 1000 trials, and mean and standard deviation are plotted in the graph. The time depends

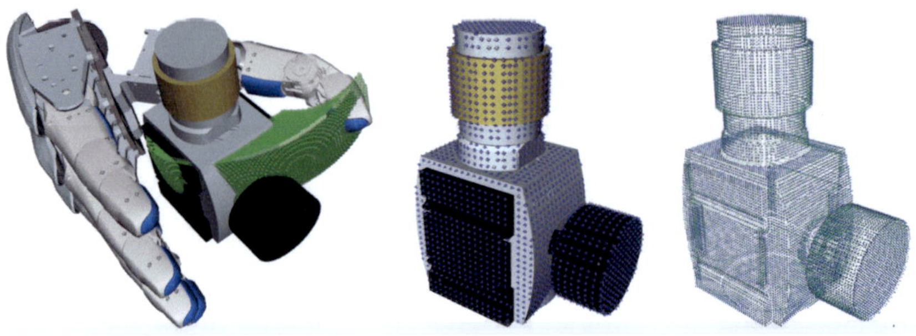

(a) (b) (c)

Figure 3.10.: Setup for the evaluation of the computation of reachable points: a) Fixed hand pose of the DLR-HIT Hand II relative to the object. The workspace of the thumb is displayed in green; b) Object discretized with $N = 3104$ points (a point every 3.5 mm) showing both the mesh and the surface points; c) Object discretized with $N = 18350$ points (a point every 1.5 mm).

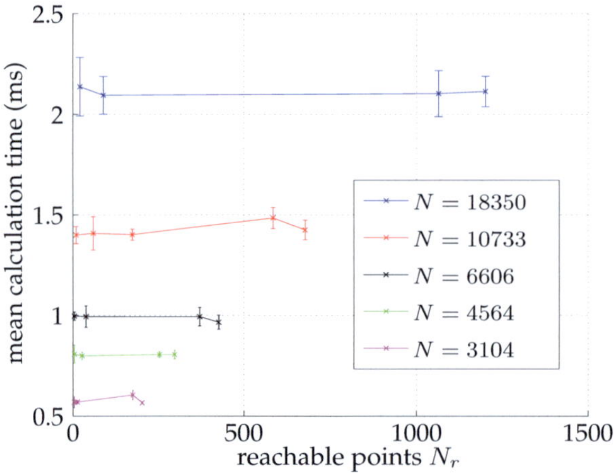

Figure 3.11.: Computational time. The colors correspond to the number N of points that discretize the object.

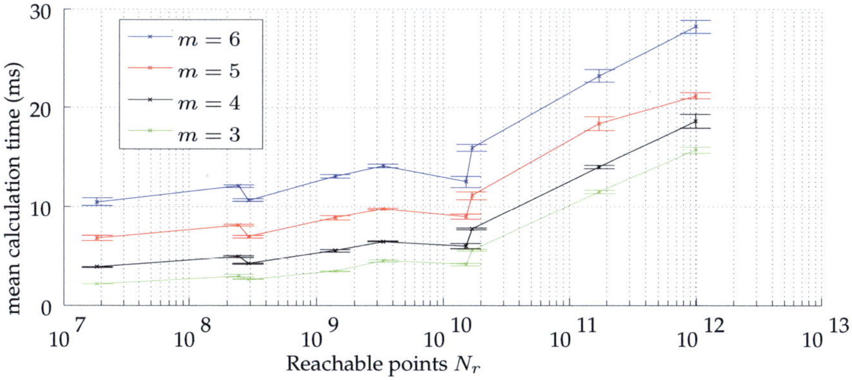

(a) 4-finger grasp

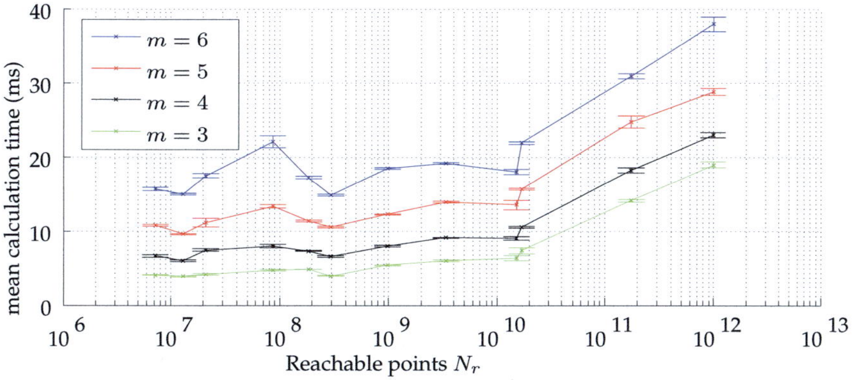

(b) 5-finger grasp

Figure 3.12.: Evaluation of the computation of rICRs using: a) 4-finger grasp; b) 5-finger grasp. The number m of sides to approximate the friction cone changes between 3 and 6.

mainly on the discretization level of the object, as it defines the total number of points that VPS must process in one query. The response time for reachability queries is under about 2 ms, even for objects with a dense point cloud.

The main factors influencing the calculation of rICRs are the number of fingers with reachable points (it could happen that not all the finger workspaces intersect the object, i.e. some $\psi_i = \emptyset$), the overall number of reachable points N_r in ψ, and the number m of sides that approximate the friction cone. Figure 3.12a and 3.12b show how the computational time changes for 4- and 5-finger grasps with respect to N_r. Clearly, the larger the number of points that the algorithm has to process, the higher the computational time. The curves do not increase monotonically, since the final time required to compute the rICRs depends also on the particular distribution of wrenches in the wrench space. Note that the complexity of the real object described with the set Ω has no influence on the computational times, as the proposed approach deals only with the point coordinates and the corresponding normals. Also, the figure shows the influence of the number m of sides for the approximated friction cone. Choosing a low m provides a very rough approximation to the friction cone, but significantly speeds up the computation; for instance, for the same number of reachable points, the computation with $m = 3$ is up to 4 times faster than with $m = 6$. The influence of m decreases in importance for a higher number of reachable points; for the figure, with the highest number of reachable points the gain in speed using $m = 3$ with respect to $m = 6$ is of 2X. For practical purposes, there must be a trade-off between computational time and precision required for the resultant rICR.

3.6. Feasible Independent Contact Regions

In this section, we present a method to plan both contact points and contact forces simultaneously. The approach is based on a ray-shooting approach presented by Zheng [231], which is introduced in Section 3.6.1.

First, we generalize the ray-shooting algorithm of [231] to consider the achievable forces by the robotic fingers. Thus, we overcome the limitation of assuming normalized contact forces. Additionally, contact forces can be computed such that they lie within the friction cone and not only on its boundary. The achievable forces are the contact forces that the finger can apply within the friction cone on the object surface given its joint angle configuration and taking the torque limits into account. Hence, these forces reflect the physical limitations of the robotic hand. Given the direction of an external disturbance wrench, for example the direction of gravity, the algorithm presented is used to calculate the maximum magnitude of the wrench that can be counteracted by the grasp with the robotic hand. That means, for example the maximum weight of the object can be computed and also updated during the execution of a task. This allows the robot to predict grasp failures due to movements of the grasped object, or to define online the limitations of manipulation actions on the grasped object. We also give a simple 2D example comparing the results of the proposed algorithm with the algorithm of [231].

Second, an algorithm to find a grasp within the reachable ICRs, which can best counteract a given disturbance wrench, is presented in Section 3.6.3. This allows us to plan contact points combined with contact forces and thus, select the grasp where the fingers can apply the maximal contact force in their current configuration. Similar to the calculation of rICRs, the algorithm is separated in an offline and an online stage, where the offline process includes the computation of the workspace of the robotic hand as previously described in Section 3.4. This workspace is then used online to obtain the physically achievable forces of the robotic hand at each potential contact point as shown in Section 3.6.2. Then, a local optimization finds the most suitable grasp. Finally, both algorithms are evaluated in Section 3.6.4.

3.6.1. Planning of Contact Forces

Zheng [231] presented a very efficient ray shooting algorithm to determine the intersection point of a ray with the convex hull of a compact convex set. He uses the duality between the convex hull of a finite number of points (e.g. $CH(W)$) and a convex polyhedron (the intersection of a finite number of halfspaces) to calculate an improved grasp quality based on W without computing the Minkowski sum explicitely and to solve the grasp force optimization problem for one specified grasp [231]. As the algorithm forms the basis for the approach proposed in this section, we provide a short introduction here; additional details can be found in [231]. First, we define the ray-shooting problem and the duality between a convex hull and a convex polyhedron.

Ray-shooting problem [126]: Let B be a given set of points $\mathbf{b}_i \in \mathbb{R}^m$ and assume that the convex hull $CH(B)$ contains the origin. Given a ray

$$R = \{\lambda \mathbf{r},\ \lambda \geq 0\} \tag{3.9}$$

emanating from the origin, find the facet of $CH(B)$ that is intersected by R at the intersection point $\mathbf{s}$. In Figure 3.13, the set B and its convex hull are shown in black, the ray R in red.

Duality [126]: If the convex hull $CH(B)$ contains the origin, it can be dually transformed to a convex polyhedron $CP(B)$. That means, there exists a reversible transformation T that maps the point $\mathbf{b}_i \in \mathbb{R}^m$ to the hyperplane $H(\mathbf{b}_i)$:

$$H(\mathbf{b}_i) : \mathbf{b}_i \cdot \mathbf{x} = b_{i1}x_1 + \ldots + b_{im}x_m = 1. \tag{3.10}$$

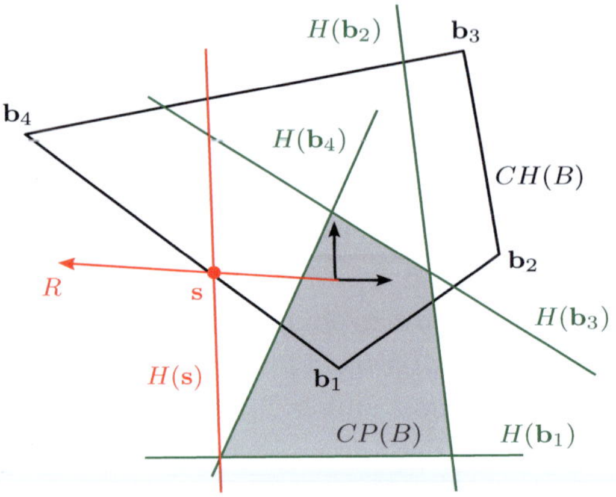

Figure 3.13.: Duality between a convex hull and a convex polyhedron: the set B and its convex hull $CH(B)$ are shown in black, the ray R in red. The dual polyhedron $CP(B)$ is the area shaded in grey, the hyperplanes $H(\mathbf{b}_i)$ are depicted in green.

The polyhedron $CP(B)$ is then bounded by hyperplanes and is described by

$$CP(B): \ \mathbf{b}_i^T \mathbf{x} \leq 1 \quad \forall \, \mathbf{b}_i \in B. \tag{3.11}$$

In Figure 3.13, $CP(B)$ is the area shaded in grey, the hyperplanes $H(\mathbf{b}_i)$ are depicted in green. The ray-shooting problem can be reformulated to finding the hyperplane $H(\mathbf{s}) : \mathbf{s}^T\mathbf{x} = 1$. The corresponding point in $CH(B)$ is the intersection point s between ray R and the facet of $CH(B)$ (as shown in Figure 3.13).

The duality between $CH(B)$ and $CP(B)$ is used in two places in Zheng's algorithm [231]: to determine the normal of a facet and for the termination criterium. The algorithm is an iterative search for the points on the boundary of $CH(B)$ that form a facet which is intersected by the ray and the intersection point is furthermost from the origin. The algorithm is visualized in Figure 3.14 and summarized in Algorithm 5. It starts with an initial set V that consists of $m+1$ points $\mathbf{w}_0, \dots, \mathbf{w}_m$ whose convex hull forms a simplex and contains the origin. That guarantees that there is a facet of the convex hull of V that intersects with the ray. A facet is described as $\mathbf{A}_j = [\mathbf{w}_0, \dots \mathbf{w}_{j-1}, \mathbf{w}_{j+1}, \dots, \mathbf{w}_m]$. Furthermore, the point $\mathbf{r}$ can be described as a convex combination of the edge points of the facet $\mathbf{A}_j$, such that $\mathbf{c}_j = \mathbf{A}_j^{-1}\mathbf{r}$. The ray intersects with the facet if $\min(\mathbf{c}_j) \geq 0$ and the intersection point $\mathbf{s}_j$ is given as

$$\mathbf{s}_j = \frac{\mathbf{r}}{\sum_j(\mathbf{c}_j)}. \tag{3.12}$$

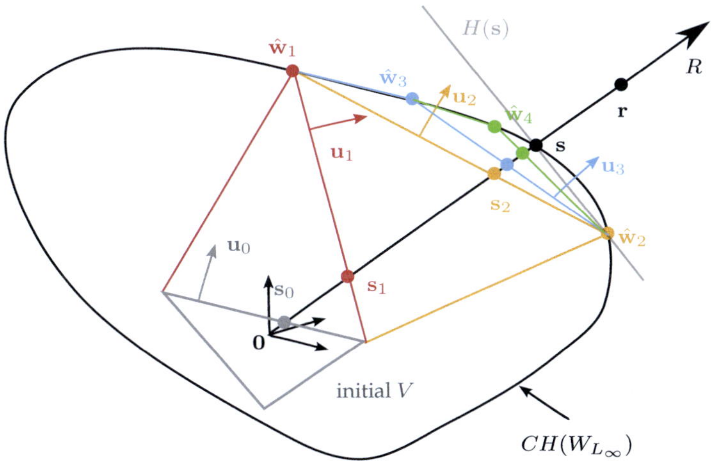

Figure 3.14.: Ray shooting algorithm proposed by Zheng [231]. The colors mark the iterations k towards the intersection point $\mathbf{s}$. In each iteration the normal $\mathbf{u}_k$ defines the search direction used to find $\hat{\mathbf{w}}_{k+1}$.

Algorithm 5: Ray shooting algorithm according to [231]

Given: Ray R through $\mathbf{r}$ and sets $W_1, W_2, \ldots, W_n$
Output: Intersection point s between the ray R and the convex hull of W_{L_∞} based on normalized contact forces

```
Form initial set V;
repeat
    j = 0;
    repeat
        j = j + 1;
        A_j = [w_0, ... w_{j-1}, w_{j+1}, ..., w_m];
        c_j = A_j^{-1} r;
    until min(c_j) ≥ 0;
    remove the jth point from V;
    compute the normal of the facet A_j (Eq. (3.14));
    add ŵ to V as its last point ( Eq. (3.18)-(3.24) );
until û − 1 ≤ ε;
return r / Σ_j (c_j);
```

Note that the norm of the intersection point s, Eq. (3.12), gives the maximum magnitude of the normalized disturbance wrench that the grasp can counteract, assuming that the contact force at each finger has a maximal normal force of one as defined by Eq. (3.2). Each of the points in $\mathbf{A}_j$ can be transformed to a hyperplane according to Eq. (3.10):

$$\mathbf{w}_j^T \mathbf{u}_j = 1. \tag{3.13}$$

Stacking the equations for all points in $\mathbf{A}_j$, the normal of the facet $\mathbf{u}_j$ is computed as

$$\mathbf{A}_j^T \mathbf{u}_j = [1, 1, \ldots, 1]^T \Rightarrow \mathbf{u}_j = (\mathbf{A}_j^T)^{-1}\mathbf{1}. \tag{3.14}$$

This normal is used as search direction to find a new point $\hat{\mathbf{w}}$ in W that has the largest projection $\hat{u}_W$ of the convex set W in direction $\mathbf{u} \in \mathbb{R}^m$. For that, [231] defines a function h_W:

$$h_W(\mathbf{u}) = \hat{u}_W = \max_{\mathbf{w} \in W}(\mathbf{u}^T \mathbf{w}). \tag{3.15}$$

The vector causing the maximal projection is $\hat{\mathbf{w}} \in W$ such that $\hat{u}_W = \mathbf{u}^T \hat{\mathbf{w}}$. Points $\mathbf{w}$ that are on the plane described by the normal $\mathbf{u}$ result in $h_W(\mathbf{u}) = 1$, points that are further from the origin result in $h_W(\mathbf{u}) > 1$. The point $\hat{\mathbf{w}}$ is added to the set V as its last point and replaces the previous jth point in the set. The procedure is repeated until $\hat{u}_W - 1 = \mathbf{u}^T \hat{\mathbf{w}} - 1 < \epsilon$. Then, the algorithm stops as the new point is on the plane described by the normal $\mathbf{u}$ (within the termination tolerance ϵ).

As computing W might be computationally expensive, we use the following Lemma to determine $\hat{u}_W$ and its corresponding $\hat{\mathbf{w}}$.
Lemma 1 The Minkowski sum of finitely many polytopes is the convex hull of the sum of its extreme points [155].

Therefore, the value $\hat{u}_W$ of the maximum projection in a certain direction in the convex set W can be calculated as the maximum of the Minkowski sum of its subsets $W_1, W_2, \ldots, W_n$. We define a set $\mathbf{v}_u$ that holds all combinations of $\hat{u}_i$ and a set $\mathbf{v}_w$ that holds all combinations of $\hat{\mathbf{w}}_i$:

$$\begin{aligned} \mathbf{v}_u &= \left\{\hat{u}_{W_1} \oplus \hat{u}_{W_2} \oplus \ldots \oplus \hat{u}_{W_n}\right\} \\ &= \Big\{\hat{u}_{W_1}, \hat{u}_{W_2}, \ldots, \hat{u}_{W_n}, \\ &\qquad \hat{u}_{W_1} + \hat{u}_{W_2}, \ldots, \hat{u}_{W_1} + \hat{u}_{W_n}, \ldots \\ &\qquad \hat{u}_{W_1} + \hat{u}_{W_2} + \ldots + \hat{u}_{W_n}\Big\}. \end{aligned} \tag{3.16}$$

$$\mathbf{v}_w = \left\{\hat{\mathbf{w}}_1 \oplus \hat{\mathbf{w}}_2 \oplus \ldots \oplus \hat{\mathbf{w}}_n\right\}. \tag{3.17}$$

The maximum projection $\hat{u}_W$ is found as

$$\hat{u}_W = \max(\mathbf{v}_u). \tag{3.18}$$

Now, $\hat{u}_W = \mathbf{v}_u(l)$ holds, and the index l is used to get the corresponding vector $\hat{\mathbf{w}}$:

$$\hat{\mathbf{w}} = \mathbf{v}_w(l). \tag{3.19}$$

Given a search direction $\mathbf{u}_j$, the calculation of $\hat{\mathbf{w}}_i$ and $\hat{u}_{W_i}$ is given in the following equations for the PCwF model; other contact models can be also used [230]. First, $\mathbf{u}_j$ is transformed into the coordinate system of the contact point

$$\mathbf{f}_i = \mathbf{G}_i^T \cdot \mathbf{u}_j. \tag{3.20}$$

According to Eq. (3.2), the maximum projection $\hat{u}_i$ of $\mathbf{f}_i$ onto the boundary of the friction cone is

$$h_T = \mu\sqrt{f_{i,2}^2 + f_{i,3}^2} \tag{3.21}$$

$$\hat{u}_i = f_{i,1} + h_T \tag{3.22}$$

with the corresponding vector

$$\hat{\mathbf{f}}_i = [1, (\mu^2 f_{i,2})/h_T, (\mu^2 f_{i,3})/h_T, 0]^T. \tag{3.23}$$

This vector always lies on the boundary of the friction. Finally, the force is transformed to the object coordinate system

$$\hat{\mathbf{w}}_i = \mathbf{G}_i \cdot \hat{\mathbf{f}}_i. \tag{3.24}$$

The overall maximum projection $\hat{u}$ and its wrench vector $\hat{\mathbf{w}}$ are calculated using Eq. (3.18) and (3.19).

3.6.2. Computation of Physically Achievable Contact Forces

In this section, the ray-shooting algorithm is generalized such that it can include physically achievable contact forces taking into account the torque limits of the robotic hand. Hence, a realistic maximum magnitude of the disturbance wrench can be computed by detecting the intersection point of the expected direction of wrench perturbation with the boundary of the convex hull of W_{L_∞}.

The mapping defined in [231] (Eq. (3.20)-(3.24)) gets the maximum projection of a direction $\mathbf{u}$ onto a force in the friction cone. Assuming that all forces in the

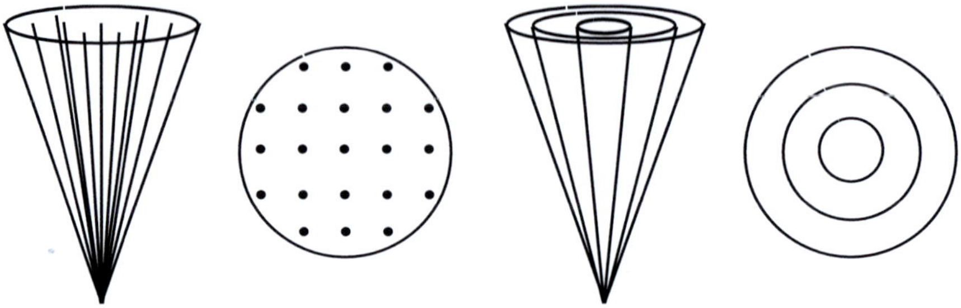

Figure 3.15.: (a) Sampled friction cone with 21 samples. (b) Three rings of friction cones with friction coefficients $\mu_1 = 0.1$, $\mu_2 = 0.2$, $\mu_3 = 0.3$.

friction cone have the same magnitude (i.e. one), this force always lies on the boundary of the friction cone. In realistic friction cones, the magnitudes of the forces might change and the maximum projection of $\mathbf{u}$ might now lie within the friction cone. There are several methods to obtain forces within the friction cone, e.g. sampling the friction cone as shown in Figure 3.15a or creating rings of cones within the original friction cone as shown in Figure 3.15b. Sampling has the disadvantage that many samples are needed which negatively affects the computation time of the algorithm. In the following, we use rings of cones to approximate the structure of a friction cone: The friction cone is divided into k friction cones with μ_k their corresponding friction coefficient. It holds $0 < \mu_k \leq \mu$. For each cone k and each contact point i, the maximum projection $\hat{u}_{i,k}$ and the vector $\hat{\mathbf{f}}_{i,k}$ with the maximal projection in direction $\mathbf{u}$ are calculated

$$\mathbf{f}_i = \mathbf{G}_i^T \cdot \mathbf{u}_j, \tag{3.25}$$

$$h_T = \mu_k \sqrt{f_{i,2}^2 + f_{i,3}^2} \tag{3.26}$$

$$\hat{u}'_{i,k} = f_{i,1} + h_T \tag{3.27}$$

$$\hat{\mathbf{f}}'_{i,k} = [1, (\mu_k^2 f_{i,2})/h_T, (\mu_k^2 f_{i,3})/h_T, 0]. \tag{3.28}$$

The force $\hat{\mathbf{f}}_{i,k}$ lies on the boundary of each normalized friction cone k. Therefore, the magnitude $\eta_{i,k}$ of this force needs to be computed according to the achievable force on this contact point i.

In this section, the corresponding joint configuration $\mathbf{q}_i$ for the contact point on the object surface is given. Hence, for each finger i, the joint torques needed to apply a force $\mathfrak{f}_i$ can be calculated with the corresponding body Jacobian $\mathbf{J}_i(\mathbf{q_i})$ [168]:

$$\boldsymbol{\tau}_i = \mathbf{J}_i(\mathbf{q}_i)^T \cdot \mathfrak{f}_i. \tag{3.29}$$

The $\mathfrak{f}_i$ marks a force described in the finger base coordinate system. The maximal magnitude η_i of the force $\mathfrak{f}_i$ is obtained by scaling the linear equation such that at least one entry of $\boldsymbol{\tau}'_i$ is equal to the corresponding torque limit ($\boldsymbol{\tau}_i$ is the vector of joint torques for finger i):

$$\boldsymbol{\tau}'_i = a \cdot \boldsymbol{\tau}_i, \text{ s.t. } \frac{\boldsymbol{\tau}'_i(k)}{\boldsymbol{\tau}_{i,\max}(k)} = 1. \tag{3.30}$$

The maximal magnitude η_i is then the magnitude of the scaled force along the normal direction, i.e. $\eta_i = a \cdot \mathfrak{f}_i(1)$.

Both, the force and the value of the maximum projection are scaled:

$$\hat{u}_{i,k} = \eta_{i,k}\hat{u}'_{i,k} \tag{3.31}$$

$$\hat{\mathbf{f}}_{i,k} = \eta_{i,k}\hat{\mathbf{f}}'_{i,k}. \tag{3.32}$$

Finally, the maximal projection and the corresponding vector are chosen according to

$$\hat{u}_i = \max_k(\hat{u}_{i,k}). \tag{3.33}$$

Formally, the new Algorithm 6 is similar to Algorithm 5, but due to Eq. (3.32) the magnitude of the contact force is shaped according to the physically achievable forces of the finger, without the linearization of the friction cone.

Example

In the following, we discuss a simple 2D example to show the influence of non-normalized forces on grasp force optimization. An ellipse is grasped with two fingers, the contacts are modeled as point contacts with friction (Figure 3.16a). The border of the friction cone (friction coefficient $\mu = 0.4$) is displayed in red in all figures. The direction of gravity is in negative y-direction: $g = [0, -1]$ (in pink). Blue marks the contact forces to lift the object with the calculated maximal weight.

We compare the calculated contact forces using 1) the original mapping by [231] (Eq. (3.18)-(3.24)) and 2) the presented mapping where scaled forces are approximated with rings of friction cones (Eq. (3.25)-(3.32)).

Algorithm 6: Ray shooting algorithm taking into account physically achievable forces

Given: Ray R through $\mathbf{r}$ and sets $W_1, W_2, \ldots, W_n$
Output: Intersection point s between the ray R and the convex hull of W_{L_∞} using physically achievable forces

Form initial set V;
repeat
    $j = 0$;
    repeat
        $j = j + 1$;
        $\mathbf{A}_j = [\mathbf{w}_0, \ldots \mathbf{w}_{j-1}, \mathbf{w}_{j+1}, \ldots, \mathbf{w}_m]$;
        $\mathbf{c}_j = \mathbf{A}_j^{-1}\mathbf{r}$;
    until $\min(\mathbf{c}_j) \geq 0$;
    remove the jth point from V;
    $\mathbf{u}_j = \mathbf{A}_j^{-T} \cdot [1, 1, \ldots, 1]^T$;
    foreach *friction cone* k **do**
        shape the wrench sets (Eq. (3.25)-(3.32));
    calculate the maximal projections $\hat{u}_i$ (Eq. (3.33));
    get $\hat{\mathbf{w}}$ (Eq. (3.18), (3.19));
    add $\hat{\mathbf{w}}$ to V as its last point;
until $\hat{u} - 1 \leq \epsilon$;
return V and $\mathbf{r}/\sum_j(\mathbf{c}_j)$;

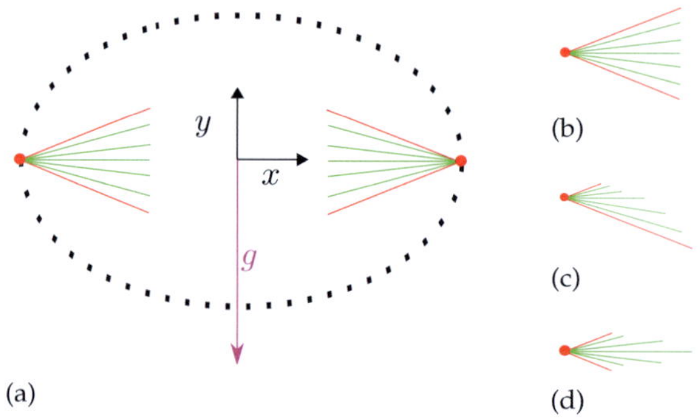

Figure 3.16.: (a) 2D example: an ellipse is grasped with two fingers. The border of the friction cone is displayed in red, the direction of gravity in pink. Particular contact forces within the friction cone with a normal component of 1 are shown in green. (b) Standard assumption: the normal component of the contact forces within the friction cone is normalized to one. (c) and (d) examples of physically achievable friction cones. The magnitude of the contact forces changes depending on the configuration of the robotic finger.

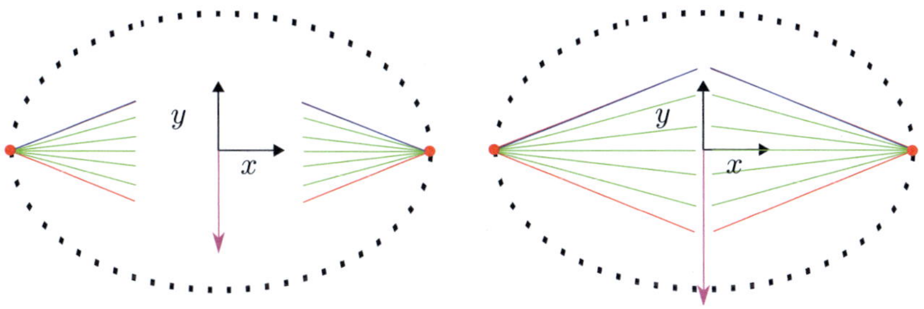

(a) (b)

Figure 3.17.: Mapping: Original ray-shooting algorithm by [231], the resulting contact forces lie on the border of the friction cones. (a) The allowed contact forces are normalized. The maximal weight of the object is 0.08 kg. (b) The normalized contact forces are scaled by 1.5. The maximal weight of the object is 0.12 kg. Scaling of the normalized contact forces does not change the direction of the resulting contact forces.

The original mapping allows only primitive contact forces at the boundary of the friction cone. Forces within the friction cone (depicted in green in Figure 3.17a) are only implicitely considered assuming that all forces have a normal component of one. Scaling these forces does not change the direction of the resulting contact force.

The next mapping function deals with non-normalized friction cones as shown in Figure 3.16c–d. Assume that the achievable contact forces for the chosen contact points on the ellipse are shaped according to Figure 3.18a. This shape is approximated with four friction cones (red, blue, orange, and green in Figure 3.18b). As the scaling of the contact forces is only dependent on the projected direction within the friction cone (Eq. (3.32)), the borders of the friction cones are scaled differently. One can see in Figure 3.18c, that the resulting contact forces move towards the center of the friction cone where the magnitude of possible contact forces is higher. With this mapping, we allow contact forces to be inside of the friction cone and not only at its border.

3.6.3. Optimal Grasp

In the previous section, we discussed the computation of a realistic magnitude of the disturbance wrench that one grasp can resist without exceeding the torque limits of the hand. This section defines an algorithm to find an optimal grasp $\mathcal{G}^{\text{opt}}$ within a set of possible force closure grasps $\mathcal{G}$, such that the fingers

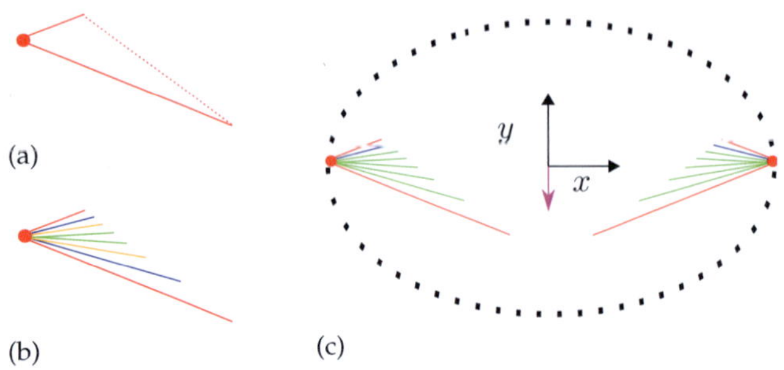

Figure 3.18.: Mapping: rings of friction cones. (a) The magnitudes of the achievable forces in the friction cone. (b) The friction cone is approximated with four friction cones with different magnitudes of contact forces. (c) The resulting contact force is now within the friction cone where the magnitude of possible contact forces is higher. The maximal weight of the object is 0.032 kg.

can counteract the largest possible perturbation along a given direction for the external wrench $\mathbf{w}_{\text{ext}}$ without exceeding their torque limits.

The set of possible force closure grasps can be obtained for example by computing reachable independent contact regions (rICRs in Section 3.5 and [251]). Reachability of the ICRs is guaranteed when the workspace of the hand is included in the computational loop [251]. However, those regions indicate reachable force closure grasps but do not provide any relation to the maximum external load that can be resisted, as the computation relies on normalized contact forces. Other grasp planning algorithms that provide sets of reachable contact points that guarantee force closure grasps are applicable as well.

The set of force closure grasps $\mathcal{G}$ consists of p grasps: $\mathcal{G}^k \in \mathcal{G}$, $k \in \{1, \ldots, p\}$. Each grasp corresponds to a set of contact points and forces with n contact points. The set of possible contact wrenches for the grasp $\mathcal{G}^k$ is defined as $W^k = W_1^k \oplus W_2^k \oplus \ldots \oplus W_n^k$.

The goal of the algorithm is to find the grasp $\mathcal{G}^{\text{opt}}$ that can counteract the largest possible perturbation without explicitly computing the largest perturbation for all grasps. We start with an initial grasp $\mathcal{G}^0 \in \mathcal{G}$ and compute the intersection point between the ray R and the convex hull of W^0 using Algorithm 6. The normal $\mathbf{u}$ of the facet intersecting with R is used to search for the next grasp $\mathcal{G}^{k+1} \in \mathcal{G}$ and its intersection point with R. The next grasp $\mathcal{G}^{k+1}$ is chosen based on the maximum projection $\hat{u}$ and its corresponding new point $\hat{\mathbf{w}}$. For that, we use Lemma 1: the value of the maximum projection $\hat{u}$ can be calculated

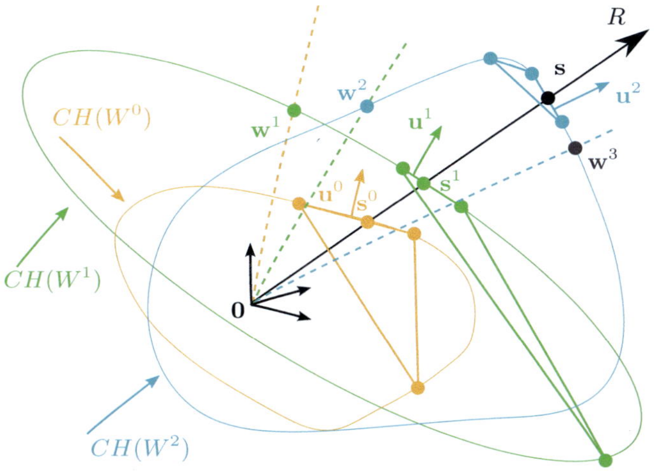

Figure 3.19.: Algorithm to find the optimal grasp $\mathcal{G}^{\text{opt}}$. Grasp $\mathcal{G}^0$ is a randomly chosen initial grasp (shown in orange). In each iteration k, the intersection point $\mathbf{s}^k$ is found using Algorithm 6. The normal $\mathbf{u}^k$ of the facet intersection with the ray R is used to find the next grasp $\mathcal{G}^{k+1}$.

as the maximum of the Minkowski sum of the maximal projections $\hat{u}_i$ for each finger i (similar to Eq. (3.18)):

$$\hat{u} = \max(\hat{u}_1 \oplus \hat{u}_2 \oplus \ldots \oplus \hat{u}_n). \tag{3.34}$$

Assume that finger i has p_i possible contact points within its reachable contact region, each one with corresponding primitive sets $W_i^1, \ldots, W_i^{p_i}$. Similar to Eq. (3.18), the maximum projection $\hat{u}_i$ is found using a set $\mathbf{v}_{u,i}$:

$$\hat{u}_i = \max(\mathbf{v}_{u,i}) = \mathbf{v}_{u,i}(l_i). \tag{3.35}$$

As a valid grasp has only one contact point per finger, the set $\mathbf{v}_{u,i}$ is defined as the union of the maximum projections of each contact point:

$$\mathbf{v}_{u,i} = \left\{\hat{u}_i^1, \hat{u}_i^2, \ldots, \hat{u}_i^{p_i}\right\}. \tag{3.36}$$

The contact point for finger i is chosen according to the index l_i. The grasp $\mathcal{G}^{k+1}$ is then defined by the indices of the contact points $l_1, \ldots, l_n$. The change between grasps is depicted in Figure 3.19.

Suppose, $\mathbf{u}^{k+1}$ points in a similar direction as $\mathbf{u}^k$, then the same grasp is found again but it does not mean that this grasp can counteract the maximum external force. To avoid that the same grasp is checked several times, we keep track of the grasps. If the grasp had already been selected, we choose a random grasp for the next iteration.

In contrast to Algorithm 6, we cannot use the duality to find a stopping criterium. With duality we can check if the new point $\hat{\mathbf{w}}^{k+1}$ is on the same dual plane as the intersection point of the previous grasp. If that is the case, $\hat{u} = 1$, but that has no relation to the distance of the intersection point. Instead, we use the distance of the intersection point as stopping criterium. As [231] pointed out, this does not guarantee the overall maximum distance of the intersection point but only a local minimum.

The complete algorithm is summarized in Algorithm 7.

Algorithm 7: Find an optimal grasp $\mathcal{G}^{\text{opt}}$ that can best resist a given wrench

Given: Ray R through $\mathbf{r}$, a set of force closure grasps $\mathcal{G}$, sets $W_1, W_2, \ldots, W_l$, and the external wrench direction.

Output: An optimal grasp $\mathcal{G}^{\text{opt}}$ described by contact points and contact forces.

Choose random grasp $\mathcal{G}^0 \in \mathcal{G}$;
$k = 0$;
repeat
    $d_{\text{old}} = d_{\text{global}}$;
    add $\mathcal{G}^k$ to the visited grasps U;
    get $\mathbf{G}$ and $\mathbf{J}$ that correspond to $\mathcal{G}^k$;
    get V^k and $\mathbf{r}/\sum_j(\mathbf{c}_j)$ by computing Algorithm 6;
    $d = [d, \mathbf{r}/\sum_j(\mathbf{c}_j)]$;
    $d_{\text{global}} = \max(d)$;
    Find new grasp $\mathcal{G}^{k+1}$ (see Eq. (3.35));
    if $\mathcal{G}^{k+1} \in U$ **then**
        choose random contact points for $\mathcal{G}^{k+1}$;
    $k = k + 1$;
until $|d_{global} - d_{old}| < \epsilon$;
return $\mathcal{G}^k$, V^k, d_{global}

3.6.4. Evaluation

We evaluate the algorithm using the DLR-HIT Hand II [124] as robotic hand. It is a five-fingered, anthropomorphic hand with modular fingers. Each finger has three degrees of freedom and all joints have the same motors and torque limits of $\tau_{\text{min}} = -1\,\text{Nm}$ and $\tau_{\text{max}} = 1\,\text{Nm}$. A particularity of the anthropomorphic design is the opposing thumb that leads to an asymmetric distribution of possible contact forces. During grasping, the thumb needs to counteract the forces of all other fingers, the external wrench, and guarantee a firm grasp on the object. Therefore, it is often the critical finger for a stable grasp.

For all experiments, the external wrench represents the gravity force of the object $\mathbf{w}_{\text{ext}} = m_o \cdot 9.81\,{}^{\text{m}}/_{\text{s}^2} \cdot \mathbf{g}$, where m_o is the weight of the object and g the normalized direction of gravity. Note that the computation of the rICRs does not depend on g. The contact regions only depend on the relative pose between hand and object.

The given computation times are based on a implementation using Matlab/Simulink and run on a standard Linux computer with an Intel(R) Xeon(R) CPU E5 - 1620 v2 processor running at 3.7GHz.

In a first experiment, we compute the achievable contact forces using Algorithm 6 given one grasp to show the influence of the torque limits of the robotic hand on the computed contact forces. Then, Algorithm 7 is evaluated showing the advantage of simultaneous computation of contact points and contact forces. By choosing appropriate contact points, lifting an object becomes feasible despite the limiting strength of the finger motors.

Evaluation of Algorithm 6

As a first evaluation of the algorithm, we show the dependency of the computation time on the number of iterations, rings of friction cones, and on the number of fingers that form the grasp. We compare five different grasps where the contact forces are computed using Zheng's algorithm (Algorithm 5) and the new algorithm (Algorithm 6) with one, two, and three rings of friction cones. For each grasp, the contact forces are calculated 2000 times and mean and standard deviation of the computation times are given. Figure 3.20 summarizes the results, where red diamonds correspond to Algorithm 5, and black squares, green crosses, and blue circles represent Algorithm 6 with one, two, and three rings of friction cones. In general, the computation time rises with the number of iterations that is needed to converge to the result. This number is neither dependent on the number of fingers that form the grasp nor on the number of rings of friction cones. It rather depends on the shape of the convex hull which cannot be estimated prior to the computation. Also, the computation times of Zheng's algorithm are always slightly smaller than with the new algorithm.

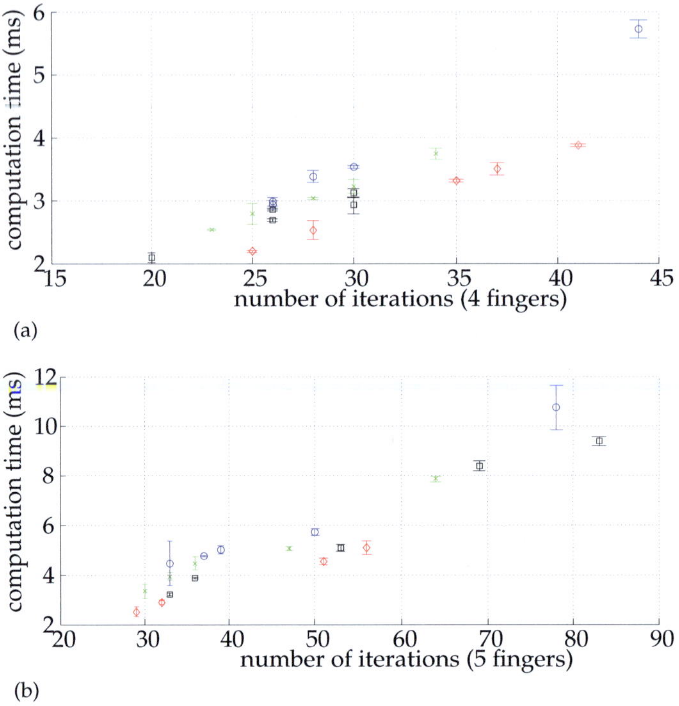

Figure 3.20.: Computation times, sorted by the number of iterations. Red diamonds: using the Zheng's mapping, black squares: one ring of friction cones, green crosses: two rings of friction cones, blue circles: three rings of friction cones (a) grasp with four fingers (b) grasp with five fingers

This results from additional computations like the Jacobian and the varying number of rings of friction cones. As can be seen in Figure 3.20b, the computation times can rise above 10 ms if the number of iterations is very high.

To analyze the computation times in detail, Figure 3.21 shows the computation time per iteration for the same 40 grasps that are shown in Figure 3.20. This time, the dependency of the computation time per iteration on the number of rings of friction cones is shown more clearly. The more rings, the higher the time starting from Zheng's mapping which stays below 0.1 ms per iteration up to three rings of friction cones which reach an average computation time of below 0.14 ms.

Given one grasp with five contact points on a box, as shown in Figure 3.22a, we evaluate the computation of achievable contact forces. The box to grasp

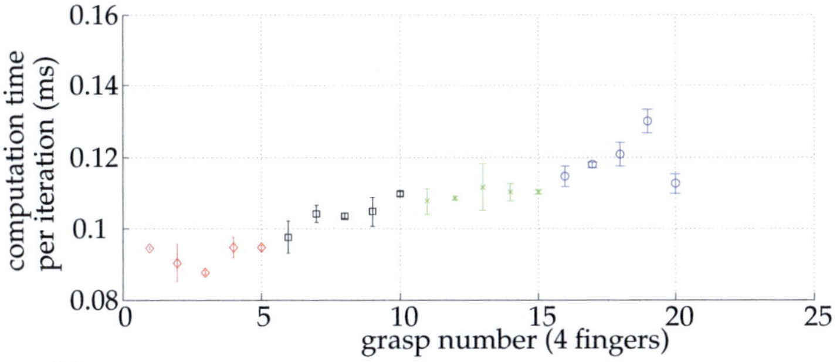

(a)

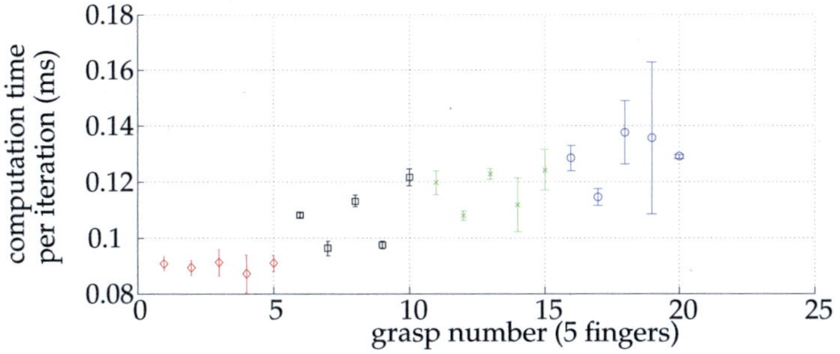

(b)

Figure 3.21.: Computation times per iteration, sorted by the number of friction cones. Red diamonds: using the Zheng's mapping, black squares: one ring of friction cones, green crosses: two rings of friction cones, blue circles: three rings of friction cones (a) grasp with four fingers (b) grasp with five fingers

weighs $m_o = 0.5\,\text{kg} \mathrel{\widehat{=}} \|\mathbf{w}_{\text{ext}}\| = 4.91\,\text{N}$. The direction of gravity is indicated with the pink arrow. Figure 3.22b shows the contact forces computed using Algorithm 5, assuming that the maximum contact force is normalized to 1 N. The calculated maximum force that the grasp can counteract is then 0.96 N. Note that the contact forces of the middle and the ring fingers are close to zero as they are not needed to counteract the external wrench. Figure 3.22b shows the contact forces for the physically achievable contact forces of the robotic finger. They are computed using Algorithm 6 with a friction coefficient of $\mu = 0.5$ and three inner friction cones. The display of the force arrows is downscaled with a factor of 0.2 due to space restrictions. The realistic maximum force that the hand can counteract is 7.70 N. Compared to Zheng's algorithm, the direction of the achievable contact forces moves towards the middle of the friction cone from 0.32 ° for the thumb to 0.001 ° for the pinky.

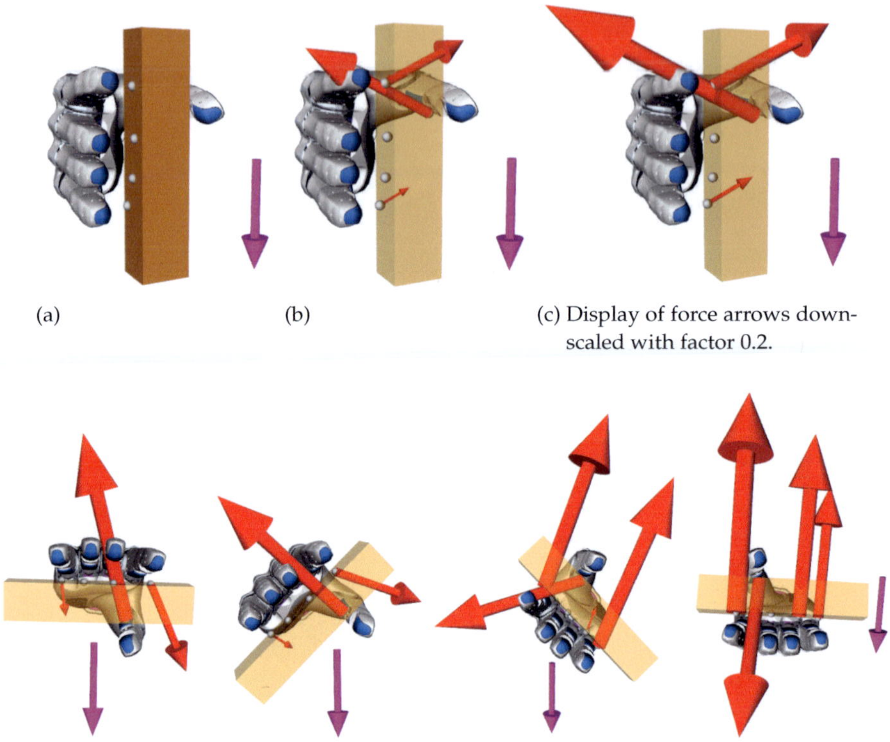

(a) (b) (c) Display of force arrows down-scaled with factor 0.2.

(d) (e) (f) (g)

Figure 3.22.: (a) Given grasp on a box. The direction of the gravity force **g** is shown in pink. (b) Using normalized contact forces of maximal 1 N, the maximum weight of the object can be 0.96 N. (c) Using the physically achievable forces of the robotic fingers, the maximal weight of the object is 7.70 N. (d) Rotation angle 90 °, maximal weight 2.49 N (e) Rotation angle 45 °, maximal weight 3.16 N (f) Rotation angle -45 °, maximal weight 19.44 N (g) Rotation angle -90 °, maximal weight 25.62 N

The computation of achievable contact forces allows monitoring the realistic maximum weight that the hand can counteract and its importance becomes obvious when the object is rotated, as shown in Figure 3.22d–g. The contact points and the hand pose relative to the object stay the same as in Figure 3.22a. During a positive rotation around the z-axis of the robotic hand, the weight of the object has to be counteracted mostly by the thumb. Its torque limits are the bottleneck for the maximum weight that can be counteracted and the magnitude of the external wrench decreases to 2.49 N for a rotation angle of 90 ° (Figure 3.22d). If the object is rotated -90 °, the weight can be counteracted

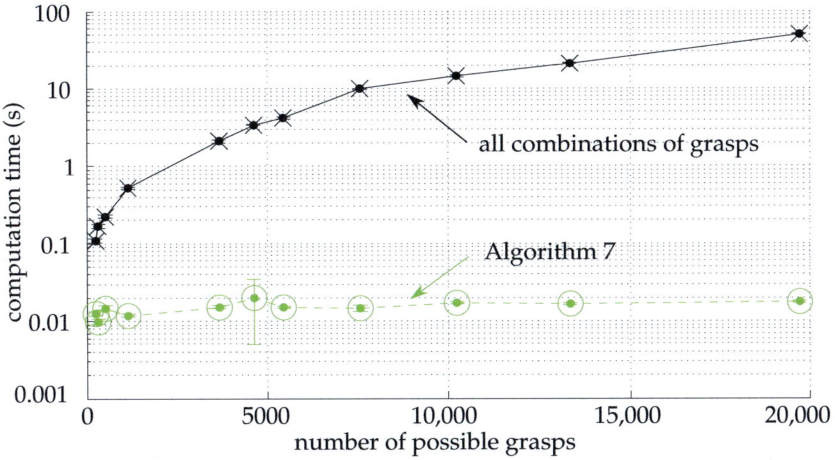

Figure 3.23.: Computation time of Algorithm 7 (green line with circles) and the calculation of all combinations of grasps (black line with crosses). The y-axis is given in log scale.

by four fingers of the hand and such, the maximum magnitude of the external force increases to 25.62 N (Figure 3.22g). For all shown grasps (Figure 3.22), the average computation time of the algorithm [231] is 3.3 ms (std 1.0 ms). The presented algorithm takes 5.5 ms on average (std 1.7 ms).

Evaluation of Algorithm 7

This section analyzes Algorithm 7, finding an optimal grasp out of a set of reachable FC grasps. We use rICRs as this set in all evaluations and the initial grasp $\mathcal{G}^0 \in \mathcal{G}$ for the search of the local optimum is formed by the contact points located in the centroid of each region. We show the computation time of the algorithm and compare it to the time needed to find the global optimum by evaluating all combinations of grasps. For different sizes of contact regions, Algorithm 7 is calculated 1000 times and the average computation time with its standard deviation is reported in Figure 3.23 indicated by the green line with circles. The y-axis is displayed in log scale. The black line with crosses visualizes the times for evaluating all combinations of grasps. Clearly, the computation time for all combinations increases with the number of possible grasps, but the computation time of Algorithm 7 stays nearly constant. It is on average 14.93 ms (standard deviation 2.74 ms) with a maximum of 0.48 s. The computation of all combinations increases from 0.10 s to 54.69 s.

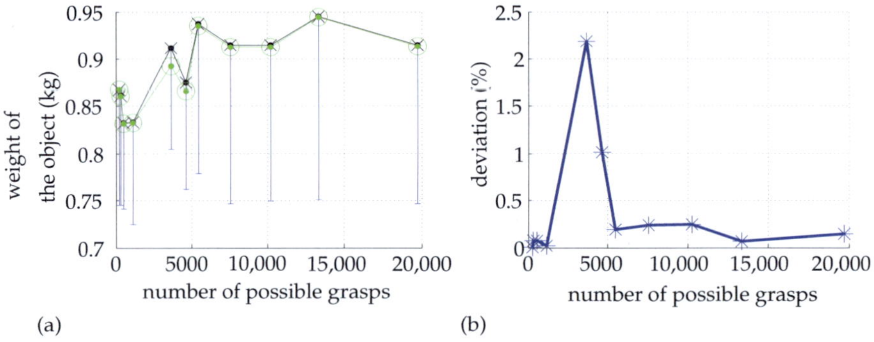

Figure 3.24.: (a) The maximal weight of the object that can be counteracted by the optimal grasp within the set of rICRs (black line with crosses). The locally optimal result of Algorithm 7 is given as the green line (with circles). The blue bars indicate the range of weights that the grasps within all possible grasps can counteract. (b) The deviation between local and global optimum is given in percentage.

Next, we compare the results achieved by computing the local optimum with Algorithm 7 compared to the global optimum, see Figure 3.24. We choose set of grasps that lead to a similar global maximal weight of the object to ensure comparable results. The optimal weight of the object varies from 0.83 kg to 0.95 kg. For the shown examples, the result of the local optimum deviates only a maximum of 2.2 % of the global result, see Figure 3.24b. This shows a clear benefit using the developed algorithm as it calculates results close to the global optimum with a low computation time.

To analyze the algorithm in detail, we use one hand pose relative to the object which allows for a three finger grasp. The setup is visualized in Figure 3.25a with a typical distribution of contact regions for each finger. Each contact region and its corresponding fingertip are indicated with the same color (thumb: red, index finger: green, middle finger: cyan), and they contain 8 contact points for the thumb and the index finger and 18 contact points for the middle finger. Figure 3.25b shows the centroid of each ICR in white. Without taking the contact forces into account, these contact points are chosen as target grasp and they represent the initial grasp for the computation of the optimal grasp. In the same figure, the contact points for the optimal grasp are visualized in black. As can be seen, they are moved towards the edges of the contact region. Clearly, the advantage for an optimal grasp must be balanced with the requirement for a robust grasp. Figure 3.25c shows the contact points for the grasp which can counteract the minimal magnitude of a disturbance wrench in the direction of gravity in white, the optimum grasp is still shown in black. The magnitude of the disturbance wrench can be 9.0 N for the white grasp and 10.0 N for the black grasp. Additionally, Figure 3.25d shows the minimal (white) and the optimum grasp (black) for a different direction of gravity, where the white grasp can counteract 1.4 N and the black grasp 3.3 N.

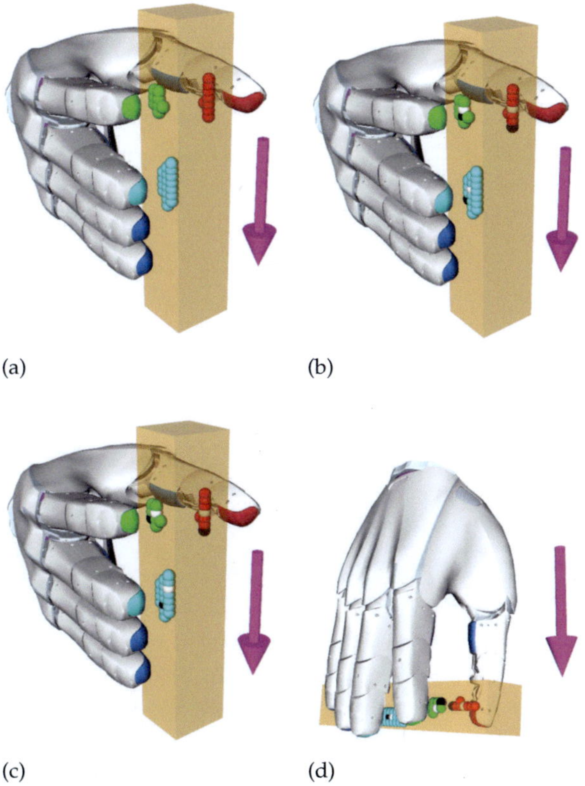

(a) (b) (c) (d)

Figure 3.25.: (a) Initial rICRs for a three finger grasp. In pink: external wrench $\mathbf{w}_{ext}$ (b) Centroid contact points of each rICR are shown in white, and the contact points for the optimal grasp in black. (c) The contact points for the grasp which can counteract the minimal magnitude of a disturbance wrench in the direction of gravity in white, the optimal grasp in black. (d) The minimal (white) and the optimal (black) grasp for a different disturbance direction

3.7. Conclusion

This chapter introduced the theoretical background and the developed methods for planning grasps in a shared autonomy setup with a high immersion of the human operator. Five requirements for such a grasp planning were identified: it needs to be general, robust, realistic, usable for multi-finger hands, and fast. We were able to develop a grasp planner that meets all requirements.

By choosing an analytic approach, the planned grasps are **general** in the sense that they can be rated according to mathematical concepts like force closure and stability. Additionally, grasps on the overall surface of the object can be computed, no training of the system is needed previous to the execution, and they are applicable to objects of any shape. **Robustness** is guaranteed by choosing the concept of independent contact regions (ICRs). This method allows to compute contact regions instead of single contact points. If the finger is placed within its corresponding region, force closure is guaranteed. Hence, instead of demanding that the robotic finger touches the object at one specific contact point, like traditional analytic grasp planning concepts, the finger only needs to be placed within the computed region. This gives inherent robustness towards finger positioning errors or uncertainties in the object detection.

In order to plan **realistic** grasps, the concept of ICRs was enhanced in this thesis in two ways. First, the workspace boundary of the robotic hand and second the torque limits of the fingers are taken into account in the grasp planning. Considering the workspace boundary leads to force closure grasps that are reachable by the robotic hand. The method was realized such that the resulting contact regions can be touched by the robotic finger and forces can be applied in the direction towards the surface of the object. However, to obtain stable grasps from force closure grasps, the contact forces need to be computed. Therefore, we developed a method to obtain realistic contact forces that respect the torque limits of the fingers. Additionally, it allows to plan contact points and contact forces simultaneously which is a considerable advantage with respect to traditional grasp planning approaches, where contact points and forces are mostly solved as independent problems. Changing the contact points according to the forces that the robotic fingers can apply at their position enables to plan grasps that can best counteract expected disturbance forces during the task.

All developed methods can be used with a **multi-finger hand** and we show the evaluation of the algorithms with a five-finger hand that performs three, four, and five-finger grasps. The computation of reachable ICRs combined with obtaining the contact forces for a certain grasp is **fast** enough ($<$50 ms) to cope with the movement speed of the human operator. Planning contact forces and contact points simultaneously slows down the computation to an overall computation time of approx. 90 ms. Nevertheless, for crucial tasks, that are usually performed with slow motions, this is still a reasonable time.

4 Guiding the Hand Pose

This chapter describes the developed assistance for positioning the robotic hand relative to the object in order to find a stable grasp. The main requirements for such an assistance in a telerobotic scenario are the following:

General: Similar to the grasp planner, the assistance for hand positioning needs to allow for a general approach direction towards the object. This ensures that the user can profit from his task knowledge without being restricted by the assistance and too few possible approach directions.

Adaptive: The operator needs the possibility to correct the chosen grasp configurations continuously during operation, to deviate from the constraints to avoid obstacles or unforseen events, and also to retract from the object if needed.

Two assistance modes are presented within this thesis that are both based on the methods described in Chapter 3. Therefore, they lead to robust grasps that can be executed in a realistic scenario. The first mode provides the operator with visual assistance and the second one uses an offline computation of a dense database of grasps to provide the operator with virtual fixtures. The background for the visual assistance was presented in Section 3.5, and thus, this chapter focuses on the assistance with virtual fixtures. First, related works regarding the computation of grasp databases and virtual fixtures are presented in Section 4.1 and the developed approach is motivated. Then, the computation of the graspability map, a dense grasp database, is detailed and evaluated in Section 4.2. This section includes content of the previous publication [253]. Section 4.3 describes the approach for adaptive virtual fixtures based on the graspability map, and the chapter is concluded in Section 4.4.

4.1. Related Works

There are two dominant approaches to solve reach and grasp tasks for autonomous robotic systems: focusing on grasping as the primary goal, and thereby considering reaching as means to the goal, or considering grasping as the final step in the reaching task. In the first approach, several solutions have been proposed in the context of autonomous manipulation to the problem of planning online the movements required to successfully grasp an object, especially when the scene is cluttered [78, 228]. In these approaches, a grasp planner is generally used to compute one feasible grasp on the object, and then the movement of the arm is planned to reach the pose of the hand provided by the grasp planner. A verification step is required at some point to determine if the planned hand pose is reachable in the current scene. If it is not reachable, a new grasp must be found. In the second approach, depending on the current motion path, it is first evaluated what hand positions and orientations can be reached in the current scenario, these are evaluated for possible valid grasps, and only then a collision-free movement of the hand/arm system is planned. For that, a compact representation of the capabilities of a robotic arm, i.e. of the different poses that the end-effector can adopt to reach different regions of the workspace, is needed like presented in [166, 226]. Such representation, called the capability map, has proved very useful in the fast planning of manipulations with a humanoid robot [229].

The grasp planner should include the hand kinematics in the computational loop, and provide a grasp configuration and the corresponding hand pose simultaneously for a particular object and hand [28, 222]. Recent approaches try to simplify the problem, either by generating sets of starting hand positions for a simplified model of the object to be grasped [70] or by using eigengrasps, i.e. hand postures obtained by a principal component analysis of the full hand configuration space [44]. The eigengrasps are used to create a grasp planner that optimizes the finger poses and wrist position and orientation to get FC grasps on the object. However, the computational times do not allow the system to be suitable for real-time applications, even for the case of a hand with low degrees of freedom such as the Barrett hand. Grasp planners could take advantage of a previous knowledge of the hand positions and orientations that might lead to an FC grasp on a particular object, therefore reducing the search space for the planning algorithm.

When the model of the object is known beforehand, such as in applications with semi-structured environments (e.g. a home environment), information useful for speeding up the online computation of a valid grasp can be obtained and stored offline, to be consulted when the online task planner requires it. For instance, the Columbia grasp database stores thousands of grasps for different objects and hands [71]. Such databases can easily be used in the second approach for grasping, as the problem of synthesizing valid grasps can be reduced to estimate the object pose and then filtering the hypotheses by reachability. Additional rankings can be applied, like the previously discussed grasp quality ε or a measure for grasp robustness, to sort the available grasps and select the one most suitable for the current scene [25]. The approaches to generate grasp databases vary mostly in the method to sample the grasp candidates, and therefore in the method to find suitable hand poses around the object. The object geometry has a large influence on the final FC grasps. Therefore, geometrical properties of basic shapes [94, 140, 141] or the axis directions of the object [171] are for example used to obtain promising hand poses. In [227], it is considered that some regions on the object surface contribute to high quality grasps more than others. This information is represented in an object-specific *grasp map*, which can later be used to bias the generation of force closure precision grasps by concentrating on the most promising regions. The construction of a *task map* for representing feasible power grasps on a particular object was previously proposed in [68]. The 6-dimensional space of positions and orientations for a particular hand with respect to an object frame is explored using Rapidly exploring Random Trees (RRTs). For a given initial pose, the hand moves forward towards the object until a contact is detected. Then, the fingers are closed until a sufficient enclosing force is achieved, and the object is lifted and slightly rotated to verify that the power grasp is successful. The exploration using RRTs allows the detection of contiguous regions of valid parameters in the pose space. The specification of continuous regions, defined as boxes in the pose space, has also been used for meeting task specifications despite of pose uncertainties [18].

Inspired by the approach to use a grasp database to choose grasps suitable for the current scenario, we propose a method to assist the operator of a shared autonomy system in positioning the robotic hand relative to the object such that a stable grasp can be executed. As the user provides the task knowledge, we consider grasping as the final step of a reaching motion, and the grasp pose should be maintained dependent on the commanded approach direction towards the object. Hence, a grasp pose is selected from a database based on functional aspects like the grasp quality ε, but also based on the user input. In order to restrict the user and his intended motion as little as possible, the database should be dense and additionally, the selection of target grasps of the database needs to be updated continuously. As assistance to reach the currently selected target grasp, the commonly used method of virtual fixtures is incorporated which also allows for providing haptic feedback for the operator [3].

Virtual fixtures constrain the movements of the user to either guide the robot along a specific pathway or prevent it from entering undesired regions [4, 188]. They have been extensively studied and applied to fields of applications like surgical or rehabilitation robotics, as summarized in the survey [31]. Several types of guidance constraints have been analyzed which can be distinguished on the dimensionality of the space they constrain. Using for example point constraints, the robot is attracted to a single point in its workspace which can be useful to assist with tool positioning [167]. Line constraints align the robot along straight lines [197] or parametric curves [132] which adds flexibility and the possibility to describe complex tool paths at the cost of higher computational effort. Furthermore, hyperplanar, parametric surface, polygonal mesh, or point cloud constraints can be incoorporated [31]. However, the vast majority of these methods considers a static constraint geometry and hence, the constraint does not change which results in inflexibility to react to changes in task or the environment. In comparison, various methods propose adaptive constraints that change depending on the user commands or the task, for example using a hidden markov model (HMM) in combination with a support vector machine to generate a line constraint composed of several straight lines [2]. Similarly, also curved constraints can be considered within this framework [225]. Although these methods show flexibility to changing environments, generating constraints with HMMs requires always training where an expert user needs to perform the intended motion several times. Another approach is to generate a library of geometrical constraints and force fields based on primitive shapes [115]. Then, these fixtures can be combined into new fixtures but have to be statically pre-defined for each task. Additionally, the introduced methods might run into problems when unpredicted obstacles or unplanned targets prevent the robot from moving along the learnt or the specified constraint. If a stiff guidance constraint is implemented, which strongly influences the motion of the robot, the user might not be able to deviate from the constraint path.

This chapter describes the novel approach of using a grasp database to create adaptive virtual fixtures to assist a user in hand positioning of a telerobotic system. Such an assistance has two main requirements: it should be general and adapative. We realize these demands in the proposed approach by presenting the computation of a dense database of grasps, the *graspability map* (Section 4.2) and by incorporating adaptive virtual fixtures (Section 4.3). The graspability map is based on the concepts of reachable independent contact regions that were introduced in the previous chapter. It represents for a particular object the positions and orientations of a mechanical hand that lead to a force closure precision grasp. Besides the potential application of graspability maps in online grasp planning, the same maps can be used to compare the precision grasp capabilities for different designs of mechanical hands. Previous attempts to compare different hands are mainly based on physical characteristics (number of fingers, number of degrees of freedom, type and number of actuators), or features related to the performance (velocity, maximum force applied at the fingertips) [6]. Two indices were presented in [20] to allow the comparison between different hands: an anthropomorphical index, which measures the similarity to the human hand in terms of aesthetic and kinematical aspects, and an index of dexterity, which tries to quantify the ability of the hand to grasp objects based on its kinematic configuration, available sensors and control system. However, aspects such as the reachable workspace of the hand and the relation of the hand structure to the objects to be grasped have not been taken into account. The graspability map considers such aspects, and therefore provides a valuable tool to compare the grasp capabilities of different mechanical hands acting on a particular object. The virtual fixtures use a simple point constraint, but provide flexibility by dynamically changing the target point dependent on the user velocity. The dense grasp database allows for smooth transitions between the changes in target poses and we additionally take into account the distance between the hand and the object to change the stiffness of the constraint. This naturally allows the user to adapt the selected grasp configuration, retract from the object, or avoid obstacles in the scene.

4.2. The Graspability Map

This section describes the approach for generating the graspability map. It is computed based on the concept of reachable independent contact regions presented in the previous chapter. We define $\Gamma = \{\mathbf{T}_0, \ldots, \mathbf{T}_m\}$ as the set of positions and orientations for the hand base frame relative to a given object where a force closure (FC) precision grasp can be achieved. Additionally, for each pose in Γ, the grasp quality ε, the size of the computed contact regions $\mathbf{z}$, as well as the finger configuration $\mathbf{q}_{\text{goal}}$ to perform the grasp are stored. The computation of the graspability map requires the following data:

- A 3D object model represented as point cloud with normals, according to Section 3.2.
- A workspace representation Φ for the fingertips of a mechanical hand, which potentially lead to FC precision grasps, as described in Section 3.4.
- A friction coefficient μ that estimates the friction between the fingertips and the object.

In order to generate the map, first, the generation of possible hand poses Γ_P relative to the given object is described and then, the computation of FC grasps and its implementation details are discussed. Although the approach to compute the graspability map is the same as in the previous publication [253], we present here several improvements to the concept:

- The computation of the FC test is not approximated, but conducted accurately in the full 6D wrench space.
- A better heuristics for the initial FC grasp is used, which more likely leads to FC grasps.
- A higher density of hand poses relative to the object is incorporated.

This section finishes with an evaluation of the graspability map and application examples.

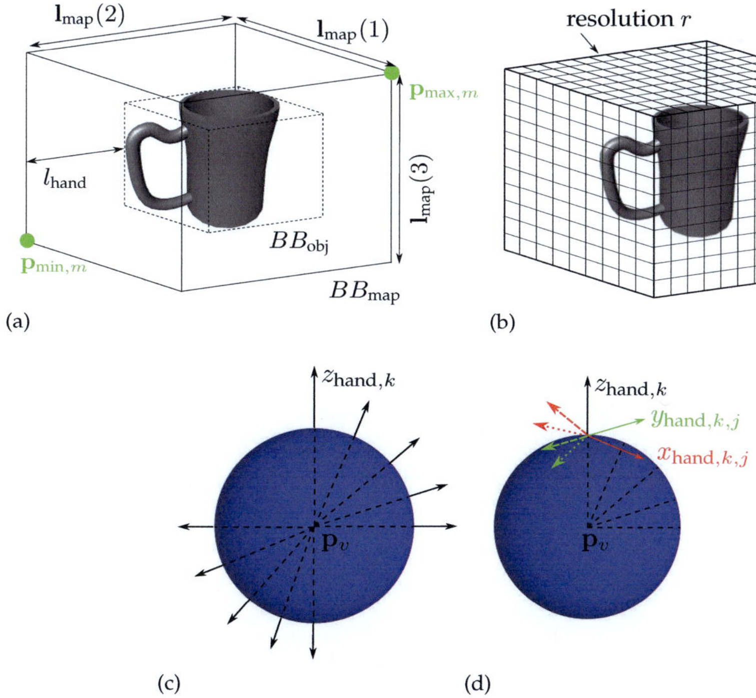

Figure 4.1.: Steps in the computation of the graspability map: a) Bounding box BB_{map} for the map, dependent on the length of the hand l_{hand} and the bounding box of the object BB_{obj}. b) Voxel grid with resolution r c) Sampling of orientations with k_z points on a sphere d) Sampling of k_{xy} roll angles around each orientation on the sphere.

4.2.1. Sampling: Generation of Hand Poses

In order to find the set Γ of hand poses that allow a valid FC grasp, we discretize $\mathbb{R}^3$ and $SO(3)$ around the object. The discretized space results in a set $\Gamma_P = \{{}^p\mathbf{T}_0, \ldots, {}^p\mathbf{T}_r\}$ of potential positions and orientations for the hand base frame, where

$$^p\mathbf{T}_i = \begin{pmatrix} \mathbf{R}_i & \mathbf{p}_{v,i} \\ \mathbf{0}^T & 1 \end{pmatrix}. \tag{4.1}$$

First, the potential positions $\mathbf{p}_{v,i} \in \mathbb{R}^3$ are obtained. The space surrounding the object that allows grasping the object with a robot hand is enveloped by a parallelepiped as bounding box BB_{map} (Figure 4.1a), which is a scaled version of the bounding box of the object BB_{obj}. Its center is the center of mass of the object $\mathbf{p}_{cm}$ and the length of its edges $\mathbf{l}_{map}$ depends on the sizes of BB_{obj} and the maximal distance from the palm to the fingertips of the robotic hand l_{hand}. A bounding box is defined by its minimal and maximal point $BB = \{\mathbf{p}_{\text{min}}; \mathbf{p}_{\text{max}}\}$, as shown in Figure 4.1a. Then, $\mathbf{l}_{\text{map}}$ can be computed as

$$\mathbf{l}_{\text{map}} = \mathbf{p}_{\text{max},o} - \mathbf{p}_{\text{min},o} + l_{\text{hand}}\mathbf{2}. \tag{4.2}$$

The bounding box is voxelized, i.e. partitioned into an axis-aligned, regular grid of cubes (Figure 4.1b), and the length of each voxel is given by the resolution r of the discretization. The number of voxels $\mathbf{n} \in \mathbb{N}$ along each axis is determined by

$$\mathbf{n} = \left\lceil \frac{\mathbf{l}_{\text{map}}}{r} \right\rceil + 1 = (\mathbf{n}(0), \mathbf{n}(1), \mathbf{n}(2))^T. \tag{4.3}$$

To obtain an equal number of voxels on each side of the center of mass, $\mathbf{n} = \mathbf{n} + 1$ for all even numbers contained in $\mathbf{n}$, and the complete number of voxels is $n_c = \prod(\mathbf{n})$. Then, BB_{map} is defined by its minimum and maximum point

$$\begin{aligned} \mathbf{p}_{\text{min},m} &= -\tfrac{\mathbf{n}-1}{2} \cdot r \\ \mathbf{p}_{\text{max},m} &= \tfrac{\mathbf{n}-1}{2} \cdot r. \end{aligned} \tag{4.4}$$

The position of each point $\mathbf{p} \in \mathbb{R}^3$ in the cartesian space within the boundaries of BB_{map} can be mapped to a unique voxel index $v \in [0, n_c - 1] \in \mathbb{N}$ by

$$\begin{aligned} \mathbf{v} &= \left\lfloor \tfrac{\mathbf{p}-\mathbf{p}_{\text{min},m}}{r} \right\rfloor = (\mathbf{v}(0), \mathbf{v}(1), \mathbf{v}(2))^T \\ v &= \lfloor \mathbf{v}(2) + \mathbf{v}(1)\mathbf{n}(2) + \mathbf{v}(0)\mathbf{n}(1)\mathbf{n}(2) \rfloor. \end{aligned} \tag{4.5}$$

Also, each voxel index v corresponds to the cartesian point in the middle of the voxel $\mathbf{p}_v$

$$\begin{aligned} \mathbf{v}(0) &= \left\lfloor \tfrac{v}{\mathbf{n}(1)\mathbf{n}(2)} \right\rfloor \\ \mathbf{v}(1) &= \left\lfloor v - \tfrac{a\mathbf{n}(1)\mathbf{n}(2)}{\mathbf{n}(2)} \right\rfloor \\ \mathbf{v}(2) &= \lfloor v - a\mathbf{n}(1)\mathbf{n}(2) + b\mathbf{n}(2) \rfloor \\ \mathbf{p}_v &= \mathbf{p}_{\text{min},m} + \tfrac{r}{2}\mathbf{1} + r\mathbf{v}. \end{aligned} \tag{4.6}$$

In a second step, the orientations in $SO(3)$ are sampled. In each voxel, a sphere is inscribed, and along its lines of latitude and longitude, k_z points are sampled by varying the two angles of the spherical coordinates (Figure 4.1c). The normal to the sphere at a sphere point determines the z-axis (black arrow) of the hand frame. Then, x- and y-axis are determined such that they are tangent to the surface of the sphere and form a right hand coordinate system. The resulting frame is rotated equidistantly around its z-axis to obtain k_{xy} frames (Figure 4.1d). Thus, each voxel v contains $k_z k_{xy}$ hand orientations, whereas the position of the base of the hand is for all of them at $\mathbf{p}_v$. Algorithm 8 formalizes the process to obtain hand poses for the computation of the graspability map.

Algorithm 8: Generate all potential hand poses Γ_P

Given:
- bounding box of the object BB_{obj}
- maximal distance from the palm to the fingertips of the robotic hand l_{hand}
- resolution r of the voxel grid
- number of discretization steps for the orientation k_z and k_{xy}

Output: $\Gamma_P = \{{}^P\mathbf{T}_0, \ldots, {}^P\mathbf{T}_r\}$

```
Obtain the bounding box of the map BB_map (Eq. (4.4));
Voxelize the enclosed space with the resolution r;
Get all positions p_{v,i} of the hand base (Eq. (4.6));
Sample a sphere uniformly with k_z points on its surface;
z-axis ← surface normal at each sampled point;
Form a right handed coordinate frame with x- and y-axis tangent to the
  surface of the sphere;
Rotate this frame k_xy-times;
forall the positions p_{v,i} do
    forall the k_z k_xy orientations per sphere do
        ^P T_i = ( R_i  p_{v,i} ; 0^T  1 );
        Γ_P.push_back(^P T_i);
```

$$ {}^P\mathbf{T}_i = \begin{pmatrix} \mathbf{R}_i & \mathbf{p}_{v,i} \\ \mathbf{0}^T & 1 \end{pmatrix}; $$

4.2.2. Algorithm

In this section, we present the approach to compute the graspability map by obtaining the hand poses $\Gamma \in \Gamma_P$. It is summarized in Algorithm 9 and uses Algorithm 10 to initialize the search for contact regions. First, all possible hand positions and orientations Γ_P are computed using Algorithm 8, described in the previous section. Then, a slightly modified version of the computation of reachable ICRs (Algorithm 4 presented in Section 3.5.1) is used to search for valid FC grasps.

Algorithm 9: Computation of the graspability map Γ

Given:

- set of points $\mathcal{P}$ describing the object surface and corresponding primitive wrenches W_i from the linearized friction cones
- set $\Phi = \bigcup_i \phi_i$ of the fingertip workspaces
- friction coefficient μ
- minimum grasp quality $\varepsilon_{\min} = \alpha\varepsilon$
- resolution of the voxel grid r
- number of discretization steps for the orientation k_z and k_{xy}

Output: Γ, ε, $\mathbf{z}$, $\mathbf{q}_{\text{goal}}$

Obtain the set of potential hand poses Γ_P; `// Algorithm 8`
foreach *potential pose* ${}^P\mathbf{T}_i \in \Gamma_P$ **do**
    if *collision between hand and object* **then**
        continue;
    Obtain random initial force closure grasp $\mathcal{C}_{\text{init}} = \{\mathbf{g}_1, \ldots, \mathbf{g}_n\}$ and reachable points ψ; `// Algorithm 10`
    Follow Algorithm 4 to compute the reachable ICRs;
    Γ.push_back(${}^P\mathbf{T}_i$);
    Store also the grasp quality ε, the size of the contact regions $\mathbf{z}$, and the goal joint configuration for the fingers $\mathbf{q}_{\text{goal}}$;

Algorithm 10: Search for a random initial FC grasp

Given:

- set of points $\mathcal{P}$ describing the object surface
- set $\Phi = \bigcup_i \phi_i$ of the fingertip workspaces

Output: $\psi = \{\psi_1, \ldots, \psi_n\}$, $\mathcal{C}_{\text{init}}$

foreach *Finger* i **do**
    Compute $\psi_i \leftarrow \phi_i \cap \mathcal{P}$;
    $n_{\psi,i} \leftarrow$ number of reachable points in ψ_i;
    if *not* $\psi_i = \emptyset$ **then**
        Get random contact point $\mathbf{g}_i \in \psi_i$;
Form the initial grasp $\mathcal{C}_{\text{init}} = \{\mathbf{g}_1, \ldots, \mathbf{g}_l\}$, and obtain the set $\mathbf{W}_{\text{init}}$ of primitive wrenches ; `/*` $\mathcal{C}_{\text{init}}$ `must contain at least` $l = 2$ `reachable points */`
Check force closure ; `/* Uses a ray-shooting test, [232] */`
cnt = 0;
while *($\mathcal{C}_{init}$ is not FC)* $\wedge$ (cnt < min(maxCnt, $\prod_i n_{\psi,i}$)) **do**
    Look for a new random grasp $\mathcal{C}_{\text{init}}$ with $\mathbf{g}_i \in \psi_i$;
    Check force closure ; `/* Uses a ray-shooting test, [232] */`
    cnt $\leftarrow$ cnt + 1;

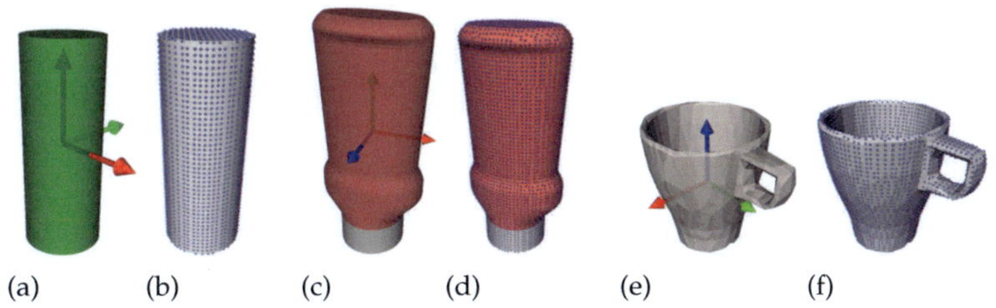

Figure 4.2.: Benchmark objects and their pointshell representation (a) a cylinder (b) cylinder sampled with 2616 points (c) a ketchup bottle (d) ketchup bottle sampled with 4578 points (e) a coffee mug (f) coffee mug sampled with 4842 points. The coordinate axis are given in RGB-order (x-axis red, y-axis green, and z-axis blue).

For each potential pose ${}^{p}\mathbf{T}_i \in \Gamma_P$, a collision detection between the object and the hand is performed, with the fingers in the configuration of maximum aperture of the hand. If there is no collision, then the intersection ψ_i between the object and the workspace ϕ_i for each finger is computed and the reachability of the points is verified according to Section 3.5. If at least two sets ψ_i are not empty (as a grasp requires at least two fingers in contact with the object), a random initial FC grasp is searched for and if one is found, rICRs are computed. The search for an initial FC grasp is capped if either a fixed maximal number of iterations maxCnt, or the amount of all possible combinations of reachable contact points $\prod_i n_{\psi,i}$ is reached. The algorithm returns not only the set Γ of hand poses that lead to a FC grasp but also the computed grasp quality ε, the size of the contact regions $\mathbf{z}$, and the goal joint configuration for the fingers $\mathbf{q}_{\text{goal}}$. In this way, one representative grasp configuration is stored for each pose that leads to a FC grasp.

4.2.3. Evaluation and Application Examples

This section presents the evaluation of the graspability map as well as some examples using three benchmark objects: a cylinder, a ketchup bottle, and a coffee mug, shown in Figure 4.2. They are represented with 2616, 4578, and 4842 points respectively. Also, we use two hands, namely the five-fingered DLR-HIT Hand II [124], and the four-fingered DLR Hand [38], to show how the graspability map can be used to compare the grasp capabilities of different robotic hands (Figure 4.3).

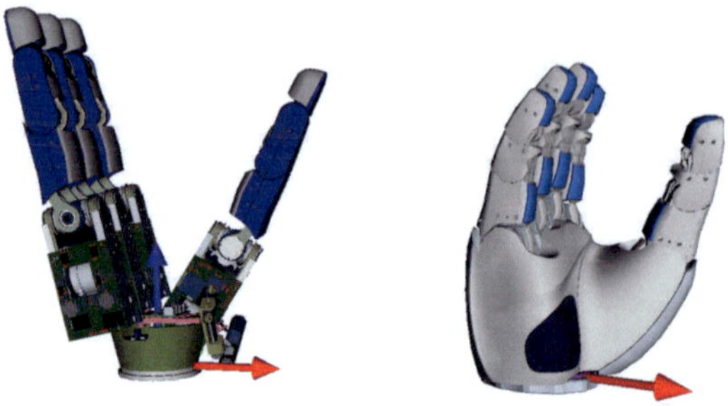

(a) (b)

Figure 4.3.: Benchmark hands (a) the four-fingered DLR Hand [38] (b) the five-fingered DLR-HIT Hand II [124]

The proposed approach was implemented in C++. An overview of the computational times is given, as evaluated on a standard Linux computer with an Intel® Xeon® CPU E5-1620 v2 processor running at 3.7 Ghz. The algorithm can be highly parallelized as the computation of each hand pose is independent of the result of the other poses. Currently, we parallelize the computation per batches of 500 voxels, but this can of course be improved. A mean of 1800 hand poses are evaluated per minute (with a standard deviation of 10 poses), and thus, the computation of 500 voxels with 1,424 poses each ($k_z = 122$, and $k_{xy} = 12$) takes approximately 6.6 h.

First, we evaluate the dependency of the map on its sampling parameters. The resolution r, and the parameters k_{xy} and k_z for the translational and orientational discretization are varied. We use the cylinder as basic shape as test object in combination with the DLR-HIT Hand II and the results are summarized in Table 4.1. The size of the bounding box of the cylinder is $(0.05, 0.05, 0.1245)$, resulting in three step sizes for the resolution r (0.01 m, 0.02 m, and 0.03 m). We compare $k_z = 122$ vs. 62 points on the sphere, and $k_{xy} = 12$ vs. 6 rotations around the z-axis of the object. Apparently, for this hand/object combination, both numbers of rotations k_{xy} and k_z are important, as decreasing their number results in half of the maximum of valid poses per voxel. Nevertheless, lowering k_z and keeping $k_{xy} = 12$ results in less decrease of valid FC poses than lowering k_{xy}. In terms of resolution r, a smaller stepsize is beneficial to reach a high number of valid grasp poses. To balance the computational effort and accuracy of the results, we choose $r = 0.02$, $k_z = 122$, and $k_{xy} = 12$ for the following graspability maps.

Table 4.1.: Variation of parameters r, k_{xy}, and k_z on the graspability map for the cylinder

resolution r [m]	0.01			0.02			0.03		
number of voxels n_c	132,055			18,125			5,491		
column	1	2	3	4	5	6	7	8	9
k_z sampled points on the sphere	122	122	62	122	122	62	122	122	62
k_{xy} rotations around z-axis	12	6	12	12	6	12	12	6	12
poses per voxel	1,464	744	744	1,464	744	744	1,464	744	744
total nr of potential poses	193,328,520	96,664,260	96,664,260	26,535,000	13,267,500	13,267,500	8,038,824	4,019,412	4,019,412
valid FC poses	107,666	59,424	66,232	13,455	8,792	9,272	4,077	2,720	2,848
nr of voxels with valid poses	29,568	26,200	28,000	3,588	3,800	3,728	1,135	1,216	1,216
maximum of valid poses per voxel	18	9	10	20	9	10	16	8	9
mean (std) of valid poses per voxel	3.64 (2.65)	2.27 (1.48)	2.37 (1.60)	3.75 (2.66)	2.31 (1.47)	2.49 (1.67)	3.59 (2.55)	2.24 (1.34)	2.34 (1.63)

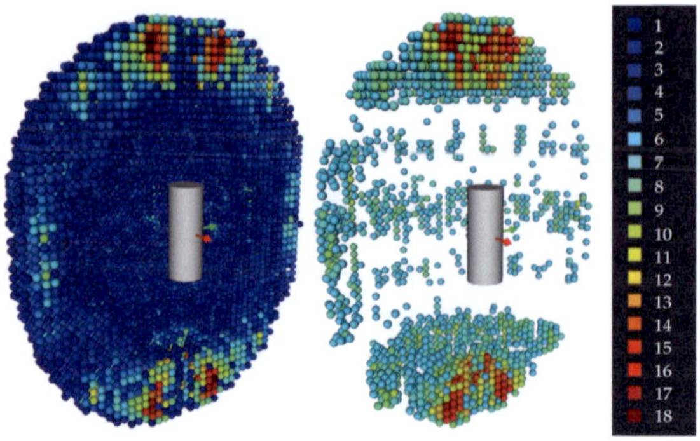

(a)

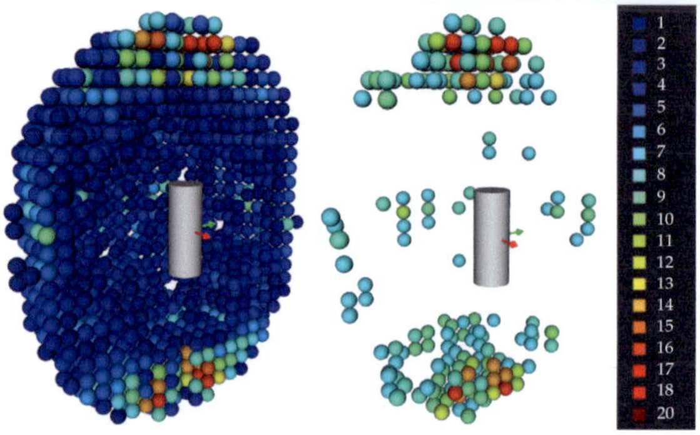

(b)

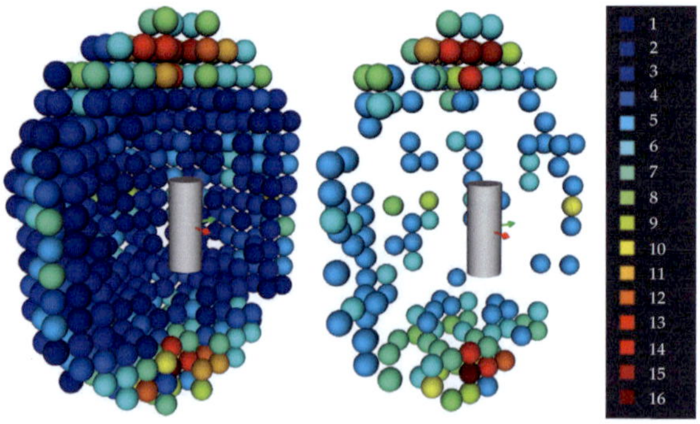

(c)

Figure 4.4.: Graspability maps for the cylinder with the five-fingered DLR-HIT Hand II. Each subfigure shows the map for all voxels with valid FC grasps on the left and on the right, only the voxels with at least 70 % of the maximal number of FC grasps per voxel are shown. a) $r = 0.01$, $k_z = 122$, $k_{xy} = 12$ b) $r = 0.02$, $k_z = 122$, $k_{xy} = 12$ c) $r = 0.03$, $k_z = 122$, $k_{xy} = 12$

Figure 4.4 shows the graspability maps for the three resolutions with $k_z = 122$, and $k_{xy} = 12$. The spheres are colored according to the number of frames on the sphere per voxel that allow an FC grasp. Red indicates regions from which the object is well graspable, and blue indicates regions from which the object can only be grasped with few hand orientations. This number, however, does not provide information about directional preferences. For clarity, we show two figures per resolution. The left one visualizes all voxel positions that contain valid FC grasps, and the right figure shows only the voxels that have a number of valid grasps which is at least 30 % of the maximum value. The colors of the spheres show how the ability of the hand to grasp an object changes around the object. One can clearly see that the regions with the highest numbers of valid grasps (voxels depicted in green to red) do not change relative to the object for different resolutions. In the case of the cylinder combined with the DLR-HIT Hand II, an approach direction from the top or bottom provides a high density of valid FC grasps.

The advantage of creating a dense database of grasps is that any chosen quality for grasps achievable at a certain voxel can be visualized on the same map. For example, the grasp quality (ε-metric) or the size of the reachable independent contact regions can be used to color-code the voxels. Then, approach directions that lead to a high grasp quality or grasps robust to position errors can be identified. For example, Figure 4.5 visualizes the graspability map for a ketchup bottle (left column) and a coffee cup (right column) for the DLR-HIT Hand II. Each row depicts a different color code: Figure 4.5a and 4.5b show the number of available FC poses per voxels, Figure 4.5c and 4.5d visualize the size of the reachable independent contact regions per voxel, and Figure 4.5e and 4.5f color code the grasp quality according to the ε-metric. Both measures are normalized: the sizes of the regions for all hand poses and the grasp quality within one voxel are separately summed and divided over the number of hand poses. Blue indicate regions that have small regions while grasp configurations in red voxels have large contact regions. Compared to the maps that visualize the number of available grasps, the robust grasps are closer to the object. In the example of the ikea mug, this suggests that grasps from the side might be more robust rather than the grasps from top. In contrast, the grasp quality shows no distinct pattern, but rather identifies single, preferred voxels. Further investigations on the consequence for grasping the object are future work.

The graspability map is also a useful tool to compare the grasp capabilities of different mechanical hands. Therefore, we use two benchmark hands: the five-finger DLR-HIT Hand II and the four-finger DLR Hand II, as shown in

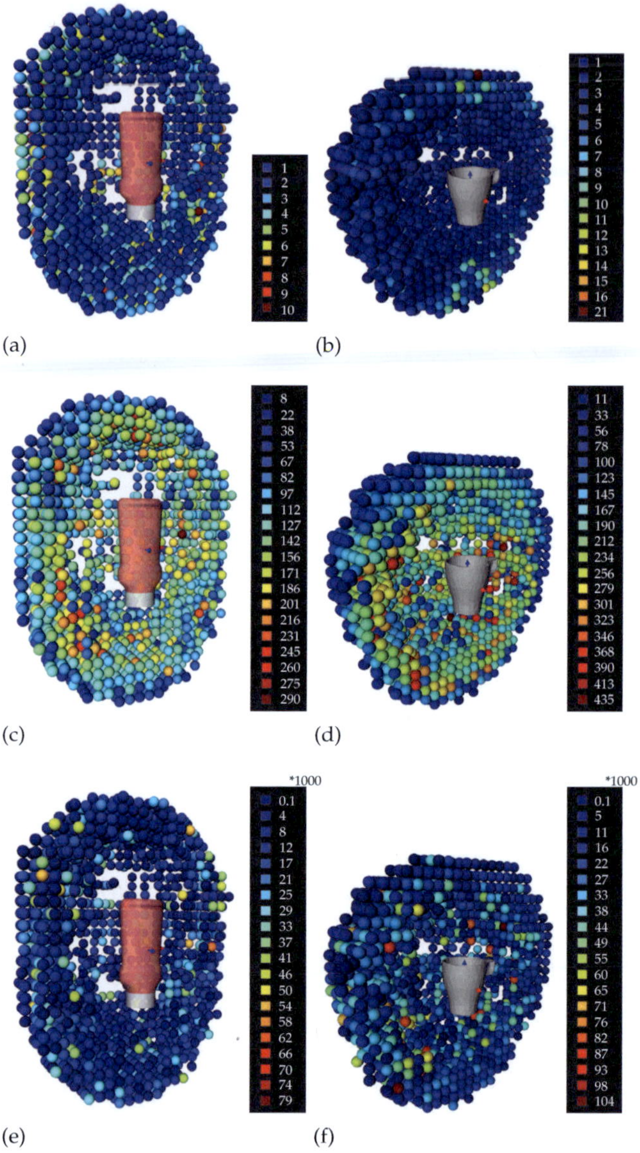

Figure 4.5.: Graspability maps for the five-fingered DLR-HIT Hand II for the ketchup bottle and the ikea mug respectively. (a) and (b) show the spheres colored according to the number of frames on the sphere that allow an FC grasp, while the coloring in (c) and (d) corresponds to the size of the reachable independent contact regions, (e) and (f) show the average grasp quality (*1000) per voxel.

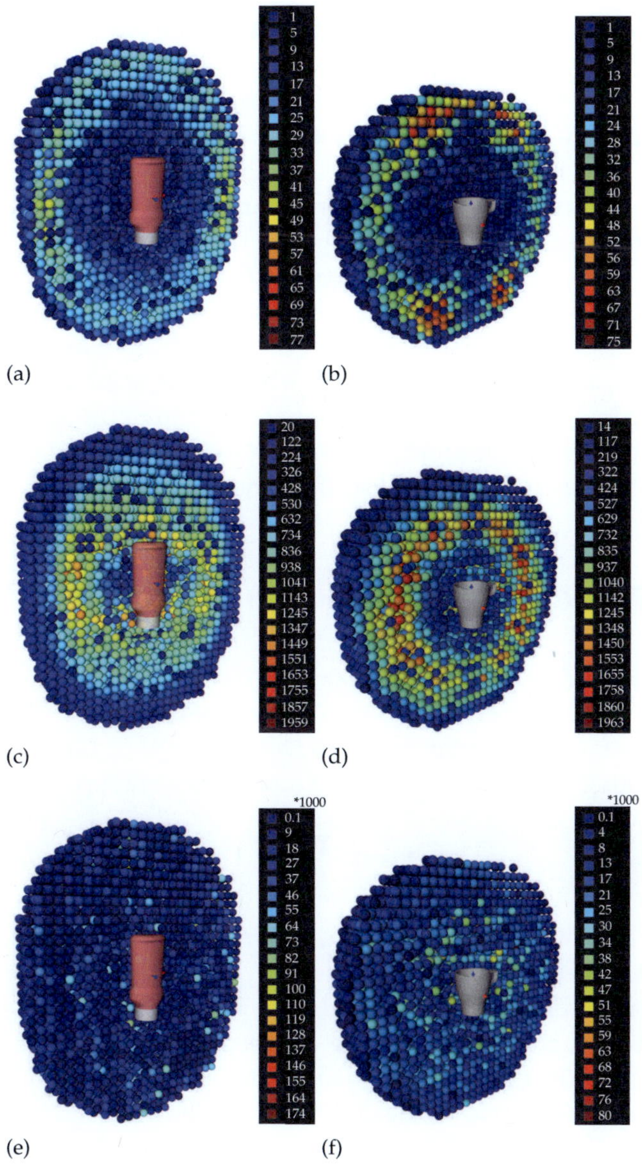

Figure 4.6.: Graspability maps for the four-finger DLR Hand II for the ketchup bottle and the ikea mug respectively. (a) and (b) show the spheres colored according to the number of frames on the sphere that allow an FC grasp, while the coloring in (c) and (d) corresponds to the size of the reachable independent contact regions. (e) and (f) show the average grasp quality (*1000) per voxel.

Figure 4.3. For instance, Figure 4.6 shows the graspability maps for the ketchup bottle and the ikea mug for the DLR Hand II. When comparing the maps for the two hands (Figure 4.5 and 4.6), the main highlight is that the graspability maps for the four-fingered hand are much larger in the size, despite the same resolution, than the other maps. In fact, that hand has a larger size, as shown in Figure 4.3, and has a larger workspace that allows more potential poses to get FC grasps on different objects. For instance, the graspability map for the ketchup bottle contains 6,964 FC grasps for the DLR-HIT Hand (Figure 4.5a) and 185,874 for the DLR Hand II (Figure 4.6a). In the case of the ikea mug (Figure 4.5b and 4.6b), although the four- and five-fingered hand are both anthropomorphic, the graspability map for the five-fingered hand contains most of its FC grasps above and below the cup. That means, this hand is able to grasp the cup using basically the internal surface of the cup or the smaller bottom, as its maximum aperture barely allows the mug to fit inside the hand, i.e. a grasp with all the fingers on the external surface is very difficult to get.

4.3. Adaptive Virtual Fixtures

This section describes the approach taken to compute adaptive virtual fixtures in order to assist the operator to reach a hand pose where a FC grasp can be executed. We follow the convention of using $\mathbf{T} \in \mathbb{R}^{4\times4}$ for a transformation matrix consisting of a rotation $\mathbf{R} \in \mathbb{R}^{3\times3}$ and a translation $\mathbf{p} \in \mathbb{R}^{3\times1}$:

$$\mathbf{T} = \begin{pmatrix} \mathbf{R} & \mathbf{p} \\ \mathbf{0} & 1 \end{pmatrix}. \tag{4.7}$$

The developed method for virtual fixtures can be divided into two major components: determining the target hand pose $\mathbf{T}_{\text{targ},k}$ and guiding the end effector, dependent on the user velocity, towards that pose. In every time step k, $\mathbf{T}_{\text{targ},k}$ is computed through an online rating of grasps within the graspability map, described in Section 4.3.1. Although all components are recomputed and updated in every time step k, we neglect the subindex k for simplicity in the following. We realize the hand guidance by influencing the translational velocity commands from the operator as well as interpolating the orientation of the end effector (Section 4.3.2). The section finishes by mentioning the implementation details. The evaluation is shown in User Study 3 in the next chapter.

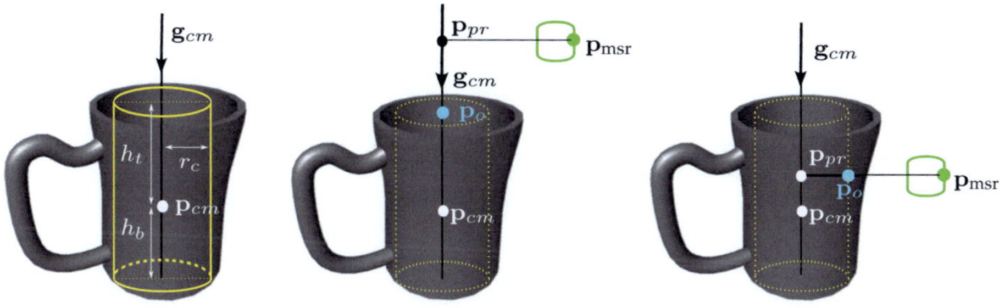

Figure 4.7.: Calculation of a point of interest $\mathbf{p}_o$ dependent on the current position of the end effector $\mathbf{p}_{\mathrm{msr}}$. a) Inscribed cylinder with radius r_c and distances h_t and h_b from the center of mass $\mathbf{p}_{cm}$ to the upper and lower intersection point of the gravity direction $\mathbf{g}_{cm}$ with the cylinder. b) The projection point $\mathbf{p}_{pr}$ is outside of the bounding box of the object and $\mathbf{p}_o$ is at the top of the object. The inscribed cylinder is depicted in yellow the c) $\mathbf{p}_{pr}$ is within the object and $\mathbf{p}_o$ is at the side of the object.

4.3.1. Online Rating of Grasps

The goal is to find a target hand pose $\mathbf{T}_{\mathrm{targ}}$ within the graspability map of the object that matches with the user intent, meaning that the resulting movement of the robotic arm and the grasp configuration are natural to the human operator. If the human chooses an approach direction above the object, we assume that it should be grasped from top, and when the object is approached from the side, the grasp pose should also be on the side of the object. Therefore, we define a point of interest $\mathbf{p}_o$ of the object that depends on the projection of the position of the current end effector pose $\mathbf{p}_{\mathrm{msr}}$ onto the direction of gravity $\mathbf{g}_{cm}$ passing through the center of mass of the object $\mathbf{p}_{cm}$. This results in the projection point $\mathbf{p}_{pr} = \Pr(\mathbf{p}_{\mathrm{msr}}, \mathbf{g}_{cm})$, visualized in Figure 4.7. Then, the point of interest is

$$\mathbf{p}_o = \begin{cases} \mathbf{p}_{pr} + r_c & \text{if } \mathbf{p}_{pr} = \Pr(\mathbf{p}_{\mathrm{msr}}, \mathbf{g}_{cm}) \text{ within the object} \\ \mathbf{p}_{cm} - \frac{h_t}{\|\mathbf{g}_{cm}\|}\mathbf{g}_{cm} & \text{if } \mathbf{p}_{pr} \text{ above the object} \\ \mathbf{p}_{cm} + \frac{h_b}{\|\mathbf{g}_{cm}\|}\mathbf{g}_{cm} & \text{if } \mathbf{p}_{pr} \text{ below the object,} \end{cases} \tag{4.8}$$

where r_c is the radius, and h_t and h_b are the distances from the center of mass to the top, resp. bottom, of the inscribed cylinder (displayed in yellow in Figure 4.7a). This cylinder is sufficient to describe $\mathbf{p}_o$ as the point of interest is only needed to determine the intended approach direction towards the object.

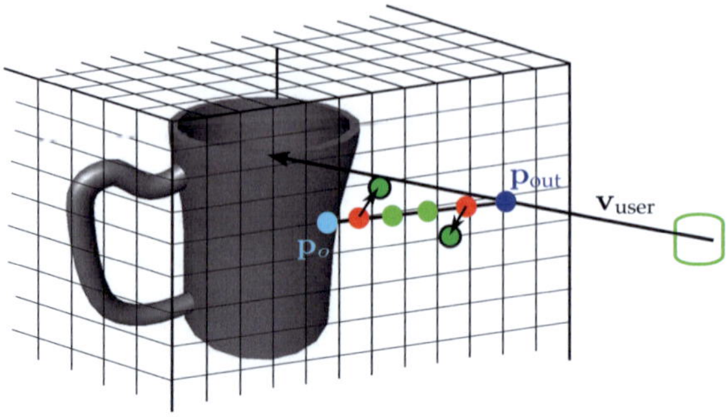

Figure 4.8.: Obtain valid hand poses from $\mathbf{v}_{\text{user}}$. Blue: intersection point $\mathbf{p}_{\text{out}}$ with BB_{map}, green: voxel with valid FC grasps, red: voxel without valid FC grasp.

Then, we use $\mathbf{p}_o$ to obtain the relevant, valid grasp poses from the graspability map in this time step: $\mathcal{T}_r \in \Gamma$, $\mathcal{T}_r = \{\mathbf{T}_{r,1}, \ldots, \mathbf{T}_{r,m}\}$. First, the intersection point $\mathbf{p}_{\text{out}}$ between the commanded velocity direction $\mathbf{v}_{\text{user}}$ and the bounding box of the map BB_{map} is found (blue circle in Figure 4.8). If the hand is already within the bounding box of the map, the intersection point $\mathbf{p}_{\text{out}}$ is not computed, but the current position of the end effector is taken instead. Then, along the line from $\mathbf{p}_{\text{out}}$ to the point of interest $\mathbf{p}_o$, all valid hand poses are obtained. The line $\mathbf{p}_o - \mathbf{p}_{\text{out}}$ is divided by the resolution r of the map and for each point on the line the corresponding voxel is found (red and green circles). For all non-valid voxels, the closest voxel with a valid FC grasp is determined (green circles with black lines, detailed in Section 4.3.3) and $\mathcal{T}_r$ contains all these valid FC grasps.

In a second step, all poses $\mathbf{T}_{r,i} \in \mathcal{T}_r$ are rated to determine the most suitable grasp configuration in the current situation. The rating depends not only on the functional aspects of the grasp, but also on how natural this pose feels for the human operator and how much the end effector needs to rotate to achieve that pose.

First, the functional aspects of each relevant grasp pose are reflected in $F(\varepsilon, \mathbf{z}_{\text{ricr}}, m_{\text{max}})_i$, which depends on the grasp quality ε, the size of the reachable independent contact regions $\mathbf{z}_{\text{ricr}}$, and the maximal magnitude m_{max} of a disturbance wrench that the grasp can counteract. Second, three parameters are used for the rating of the rotational movement of the end effector N_i: θ_{align}, θ_{tot}, and θ_{roll}. The parameter θ_{align} rates grasp configurations with respect to their alignment of the hand with the object axes. Similar to the computation of the point of interest, it depends on whether the projection of the voxel center $\mathbf{p}_{r,i}$

on the direction of gravity $\mathbf{g}_{cm}$ lies within or outside the object. The projected point is $\mathbf{p}_{pr} = \Pr(\mathbf{p}_{r,i}, \mathbf{g}_{cm})$, and the resultant major axis $\mathbf{a}$ is computed as

$$\mathbf{a} = \begin{cases} \mathbf{p}_{pr} - \mathbf{p}_{r,i} & \text{if } \mathbf{p}_{pr} = \Pr(\mathbf{p}_{r,i}, \mathbf{g}_{cm}) \text{ within the object} \\ \mathbf{p}_{cm} - \mathbf{p}_{pr} & \text{if } \mathbf{p}_{pr} \text{ outside the object.} \end{cases} \tag{4.9}$$

Then, the deviation in orientation between an axis-aligned hand and the current target pose is $\theta_{\text{align}} = \measuredangle(\mathbf{a}, \mathbf{a}_{\text{hand}}) \in [0; \pi]$, where $\mathbf{a}_{\text{hand}}$ is the axis of the hand pointing from the palm towards the fingertip, i.e. in case of the DLR-HIT Hand II the third column of $\mathbf{R}_{r,i}$ (Figure 4.9b). We rate the total angular rotation θ_{tot} between the target orientation of the grasp configuration $\mathbf{R}_{r,i}$ and the initial hand orientation $\mathbf{R}_{\text{init}}$, i.e. the start pose of the hand. Therefore, both hand poses are compared using the angle-axis representation

$$\mathbf{R}_{12} = \mathbf{R}_{\text{init}}^{-1}\mathbf{R}_{r,i} \tag{4.10}$$

$$[\beta, \mathbf{b}] = \text{angle-axis}(\mathbf{R}_{12}), \tag{4.11}$$

where $\mathbf{b}$ is the axis and β the angle of the rotation from the initial rotation to the target rotation of the hand. The total angular rotation is then

$$\theta_{\text{tot}} = \sum(\beta\mathbf{b}) \in [0; 2\pi]. \tag{4.12}$$

Third, the roll angle around the z-axis of the hand $\mathbf{a}_{\text{hand}}$ is rated specifically with $\theta_{\text{roll}} \in [0; \pi]$ using the Euler angle representation. While grasp configurations with a high rotation around this axis do not represent any problem from the grasp capability point of view, they might not seem natural, as illustrated in Figure 4.9.

In summary, the overall rating λ_i for each relevant hand pose is

$$\begin{aligned} \lambda_i &= F(\varepsilon, \mathbf{z}_{\text{ricr}}, m_{\text{max}})_i \cdot N(\theta_{\text{align}}, \theta_{\text{tot}}, \theta_{\text{roll}})_i \\ &= \exp\Big(-(f_1(\varepsilon) + f_2(\sum(\mathbf{z}_{\text{ricr}})) + f_3(m_{\text{max}}) + w_4|\theta_{\text{align}}| + w_5|\theta_{\text{tot}}| + w_6|\theta_{\text{roll}}|)\Big). \end{aligned} \tag{4.13}$$

We choose the exponential function to ensure a low weighting of grasp configurations where the evaluation of at least one of the criteria results in a low rating. To weight all criteria equally, w_j are the weights for each rating and f_1, f_2, and f_3 are functions such that the value of the exponent is normalized. Finally, the pose with the highest λ_i is selected as the target pose $\mathbf{T}_{\text{targ}}$ at that particular time step.

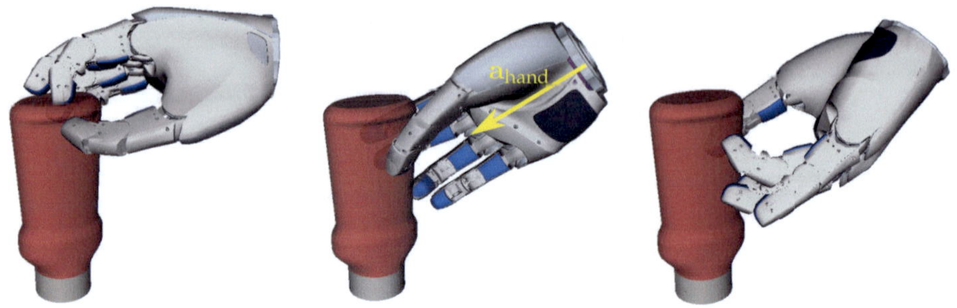

(a) (b) (c)

Figure 4.9.: Three hand poses for the same voxel are shown. While a) and b) look like natural grasps, c) seems akward. These poses are rated as $\lambda_a = 1.20$, $\lambda_b = 1.74$, $\lambda_c = 0.70$,

For the example shown in Figure 4.9, all exponents are normalized to 2π and hence, $w_4 = w_6 = 2$ and $w_5 = 1$. The normalizing functions are determined as

$$\begin{aligned} f_1 &= 2\pi^{\frac{\max(Q)-\varepsilon_i}{\max(Q)}} \\ f_2 &= 2\pi^{\frac{\max(Z)-\sum \mathbf{z}_{\mathrm{ricr},i}}{\max(Z)}} \\ f_3 &= 2\pi^{\frac{\max(M)-m_{\mathrm{max},i}}{\max(M)}}, \end{aligned} \tag{4.14}$$

where $\max(Q)$, $\max(Z)$, and $\max(M)$ are the maximal values for the grasp quality, the sum of rICR sizes, and the maximal magnitude of the disturbance wrench that the grasp can counteract. The three hand poses shown in Figure 4.9, are rated as $\lambda_a = 1.20$, $\lambda_b = 1.74$, and $\lambda_c = 0.70$.

4.3.2. Velocity Mapping

In general, the proposed method continuously modifies a commanded translational user velocity $\mathbf{v}_{\mathrm{user}}$, such that the end effector reaches a hand pose relative to the object where a stable grasp can be executed. The behavior of the algorithm depends on the distance d_{msr} between the current end effector pose $\mathbf{T}_{\mathrm{msr}}$ and the object $\mathbf{T}_{\mathrm{obj}}$: the closer the end effector is to the object, the higher the influence of the autonomy. Of course, the desired behavior depends on the size of the object as well as on the size of the mechanical hand. Therefore, we split the space around the object in four zones, as shown in Figure 4.10. In each

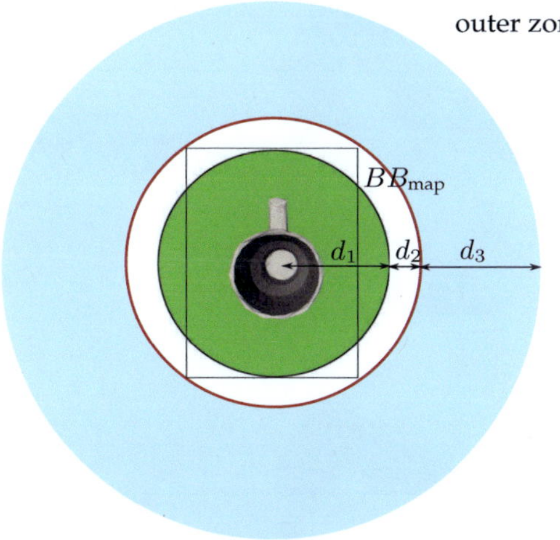

Figure 4.10.: Zones of Operation: green: autonomous zone $\alpha = 1$, white: buffer zone $\alpha = 1$, blue: interpolation zone $\alpha \in [0, 1]$, and outer zone $\alpha = 0$.

zone, the influence of the autonomy is determined by the factor $\alpha \in [0, 1]$, and the resultant velocity $\mathbf{v}_{\text{res}}$ of the end effector is

$$\mathbf{v}_{\text{res}} = (1 - \alpha)\mathbf{v}_{\text{user}} + \alpha\mathbf{v}_{\text{auto}}, \tag{4.15}$$

where the autonomous component of the velocity $\mathbf{v}_{\text{auto}}$

$$\mathbf{v}_{\text{auto}} = \|\mathbf{v}_{\text{user}}\| \frac{\mathbf{p}_{\text{targ}} - \mathbf{p}_{\text{msr}}}{\|\mathbf{p}_{\text{targ}} - \mathbf{p}_{\text{msr}}\|} \tag{4.16}$$

depends on the current position of the end effector $\mathbf{p}_{\text{msr}}$ and the position of the target hand pose $\mathbf{p}_{\text{targ}}$. Additionally, the continuous update for target hand poses $\mathbf{T}_{\text{targ}}$ also changes according to the zone. In the outer zone, where the hand is still far from the object, the autonomy has no influence on the velocity of the end effector and $\alpha = 0$. Nevertheless, when $\mathbf{v}_{\text{user}}$ points towards the object and an intersection point $\mathbf{p}_{\text{out}}$ on the bounding box of the map can be found, the target hand pose $\mathbf{T}_{\text{targ}}$ is updated. If the user retracts from the object or wants to reach towards a different goal, the search for $\mathbf{T}_{\text{targ}}$ is not carried out. As soon as the hand enters the interpolation zone ($d_1 + d_2 \leq d_{\text{msr}} \leq d_1 + d_2 + d_3$), the velocity is gradually influenced by the autonomy. In fact, α varies according to

$$\alpha = \min(e^{-(d_{\text{msr}} - d_1)/d_3}, 1). \tag{4.17}$$

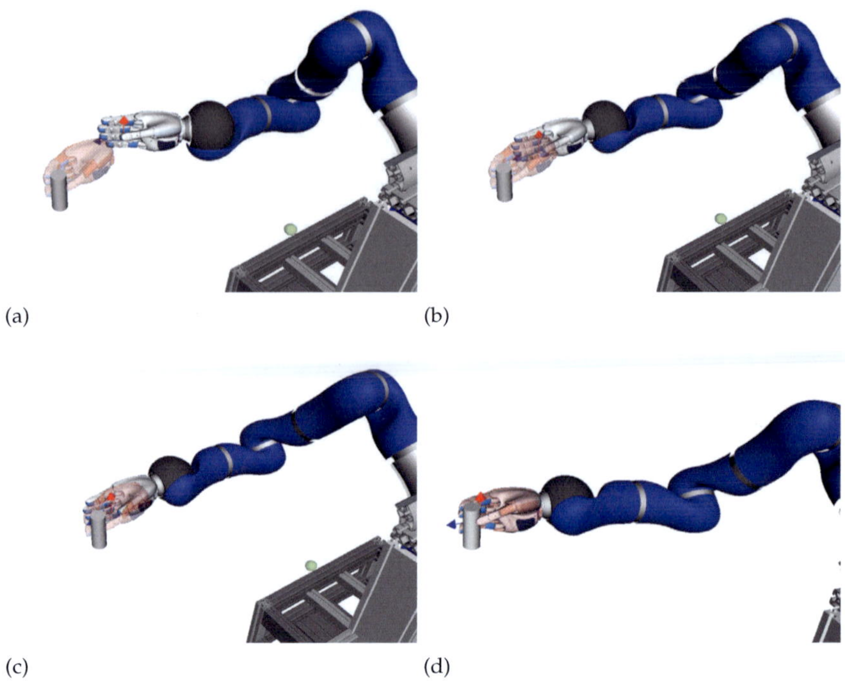

Figure 4.11.: Interpolation of the orientation a) $t = 0$, b) and c) $0 < t < 1$, d) $t = 1$.

Also, the probability of changing $\mathbf{T}_{\text{targ}}$ decays along with $1 - \alpha$. This ensures that the target pose is only updated if there is a significant improvement in the pose rating or the commanded user velocity is drastically different to previous time steps. Hence, we avoid sudden exchanges of $\mathbf{T}_{\text{targ}}$ and guarantee a smooth trajectory towards the target pose the closer the hand is to the object. Please note that this does not restrict the operator in choosing an approach direction, as he can always retract from the object. Similarly, the interpolation of the orientation of the hand pose to the orientation of the target pose is carried out as a function of the distance to the object. The interpolation fraction t is then

$$t = \begin{cases} 0 & \text{if } d_{\text{msr}} > (d_1 + d_2 + d_3) \\ (d_{\text{msr}} - d_1 - d_2)/d_3 & \text{if } d_1 + d_2 < d_{\text{msr}} \leq (d_1 + d_2 + d_3) \\ 1 & \text{if } d_{\text{msr}} \leq (d_1 + d_2). \end{cases} \tag{4.18}$$

An example is visualized in Figure 4.11. The rotation matrices for the initial and target poses are converted into quaternions (q_{init} and q_{targ}), and the intermediate quaternion q_{inter} for a given t is calculated using spherical linear interpolation (SLERP) [195]

$$q_{\text{inter}} = \frac{\sin\Big((1-t)\theta_{\text{tot}}\Big)}{\sin\theta_{\text{tot}}} q_{\text{init}} + \frac{\sin(t\theta_{\text{tot}})}{\sin\theta_{\text{tot}}} q_{\text{targ}}. \tag{4.19}$$

The distance d_3 that determines the length of this interpolation zone should be at least the length of the hand from its palm to the fingertips to enable a collision free interpolation of the hand orientation with the object. Therefore, d_3 is a tradeoff between the chosen speed of the interpolation and the radius of influence of the autonomy on the movement of the hand. When the hand enters the buffer zone, i.e. $d_{\text{msr}} = (d_1 + d_2)$, the interpolation of the orientation of the hand pose to the orientation of the target pose is completed. The buffer zone is essentially a small zone where the user can observe the hand orientation to decide if it matches his expectations and thus, if he wants to trigger a grasp or retract from the object to find a better grasp configuration. This zone enables users to make any final changes to the approach direction as the update of the target pose is still enabled. Once the autonomous zone ($d_{\text{msr}} \leq d_1$) is entered, $\alpha = 1$ and thus, the velocity direction is completely autonomous. Only if the user wants to retract, the autonomus movement towards the object is aborted. No further update of the target pose is carried out. Grasping is of course still enabled in this zone and once the target voxel is reached, the end effector cannot move further towards the object. The selected hand orientation is held and can be changed only if the end effector point moves out of the autonomous zone. Finally, if the user triggers the grasp, the end effector moves autonomously to the selected grasp point. The distance d_1 is determined by the radius of a sphere inscribed in the bounding box of the graspability map, and $d_1 + d_2$ is the radius of a sphere that circumscribes the map. Therefore, the autonomous zone lies completely within the bounds of the graspability map.

4.3.3. Implementation Details

The concept is realized by splitting the pose rating of grasps and the velocity mapping in two separate modules. This allows to execute the velocity mapping in the hard realtime of the control loop of the robot, in our case within 1 ms.

The pose rating only needs to match the commanded user velocity to provide updates to the target grasp pose without slowing down the operator. Therefore, we precompute the nearest force closure point for each voxel position that does not contain a valid force closure grasp (the red circles in Figure 4.8). As it is precomputed offline, we use a brute force search within the voxel space to search for the closest voxel (using the $\mathcal{L}_2$-norm) that contains a valid grasp configuration.

4.4. Conclusion

This chapter presents a newly developed assistance for positioning the robotic hand relative to the object in order to find a stable grasp. Two requirements need to be fulfilled: First, the operator should be constrained as little as possible in the approach directions towards the object to be able to use his task knowledge fully. Second, the human needs the possibility to correct the chosen grasp configurations continuously during operation, to deviate from the constraints to avoid obstacles or unforseen events, and also to retract from the object if needed.

We take these requirements into account by combining a grasp database with the concept of virtual fixtures to create a flexible assistance. In order to allow for a variety of approach directions, a grasp database, the *graspability map*, is developed, which covers the space around the object in a dense manner. The method of computation is based on the concept of reachable independent contact regions presented in the previous chapter, and was additionally enhanced compared to the previous publication [253]. A higher density of hand poses relative to the object is incorporated by using a better heuristics for the initial FC grasp, which leads more likely to FC grasps. The force closure test is conducted in the full 6D wrench space and thus, leads to accurate graspability maps. The parameters for choosing the discretization of sampling are evaluated, and additional use of the map for comparison of grasp capabilities of different robotic hands is discussed. The graspability map also simplifies the comparison of different quality measures of grasps like the traditional ε-metric or the size of the independent contact regions.

The virtual fixtures are adaptively created by searching through the graspability map for the most suitable grasp dependent on the commanded velocity direction of the operator. Using this dense database of grasps, the target grasp can be continuously updated according to the functional aspects of the available grasps (grasp quality, icr sizes, maximum magnitude of disturbance wrench) and the perceived naturalness of the movement of the robotic hand (overall rotation angles). As constraint geometry for the virtual fixture, a simple point constraint (the target grasp) is chosen. However, this point is not static but allows for retraction from the object and deviation from the desired path, which allows flexibility for the operator, while at the same time assisting in adopting the target grasp.

5 Evaluation of Shared Grasping

This section presents the evaluation of the developed assistance methods for grasping and manipulation with psycho-physical user studies. In all studies, objective and subjective measures are used to analyze system performance. All three challenges of shared grasping, introduced in Section 2.3, are examined:

C1 - hand positioning
: Setting up a position and orientation of the hand relative to the object.

C2 - finger adjustment
: Finding contact points and the corresponding finger joint configurations for a stable grasp.

C3 - actual grasp
: Moving into contact with all fingers, applying suitable contact forces, and ensuring grasp stability.

Three user studies are conducted that focus each on different aspects of the challenges, which is why we use two hardware setups to conduct the studies. A shared autonomy system based on DLR's telepresence system is used at first to evaluate the influence of the online grasp planning (C2) and the actual

Table 5.1.: Overview of the conducted user studies.

	Challenge	Hardware setup	End effector
Study 1	C2, C3	DLR shared autonomy system	1, 2
Study 2	C1 (visual)	DLR shared autonomy system	1, 3
Study 3	C1 (virtual fixture)	DLR assistive robotics system	1

grasp (C3) in a real scenario, where the operator experiences multi-modal feedback. This study additionally analyzes the influence of different kinematics of the end effector on the task success and the immersion of the operator. Second, the same multi-modal human-machine interface is used to study the effect of a visual assistance for hand positioning (C1) while varying again the kinematics of the end effector. Finally, a third study is conducted using a DLR's assistive robotics system to investigate the assistance for hand positioning using virtual fixtures (C1). This hardware realization is particularly suitable as the human interface in this system limits the operator in the degrees of freedom to position the hand relative to the object. Hence, the assistance is crucial to perform grasping of objects from different approach directions. Table 5.1 summarizes the conducted user studies and their specifics.

This chapter is organized as follows. The DLR shared autonomy system is introduced in Section 5.1 and the methodology for both user studies conducted on this system (Study 1 and 2) summarized in Section 5.2. The corresponding user studies are then detailed in Section 5.3 and 5.4. Section 5.5 introduces the DLR assistive robotics system which is used to analyze the assistance for hand positioning with virtual fixtures in Study 3. The user study is then presented in Section 5.6. Section 5.7 concludes the chapter with a discussion on the results of the studies.

Semi-autonomous grasping with DLR's telepresence system was previously published in [243, 257] and the first two user studies were made public in [247, 248, 259].

5.1. The DLR Shared Autonomy System

Semi-autonomous grasping is realized and evaluated using DLR's telepresence system [243]. It consists of the multimodal human machine interface HUG [249]

(a) (b)

Figure 5.1.: Semi-autonomous grasping using DLR's telepresence system. It consists of the remote robot SpaceJustin (on the left) and the human-machine interface HUG (on the right).

and the remote robot SpaceJustin, which is a modified version of DLR's humanoid robot Justin [27], see Figure 5.1a. SpaceJustin has 17 actuated degrees of freedom (DoF) for torso, head, and arms and interacts with the environment with two DLR-HIT Hands II [124]. These anthropomorphic hands have five fingers with 3 DoF each. HUG is composed of two DLR Light Weight Robot (LWR) arms with seven DoF each, which are mounted behind the user as shown in Figure 5.1b. The positions of the LWRs have been optimized to enable the operator to use the complete workspace of his arms. The robot arms of HUG and SpaceJustin are coupled in Cartesian space, which allows the operator to experience realistic force feedback on his hands. He also perceives visual feedback by wearing a head-mounted display (HMD, NVisorSX60 from NVIS[1]) showing the remote environment in 3D. Additionally, virtual scenes and visual assistances can be displayed in the HMD using the InstantPlayer from Fraunhofer IGD[2]. In order to allow a high degree of immersion, the head movements of the operator are additionally tracked so that he can directly control the movement of SpaceJustin's head by using his own. In this way, human's most effective modalities to perform prehension tasks are provided: visual feedback and haptic feedback on the palm.

For grasping, there exist two hand interfaces as shown in Figure 5.2. Cybergloves[3] allow the individual control of the fingertips of the DLR-HIT Hands while being coupled to the robot arms with a magnetic clutch (Figure 5.2a).

[1] http://www.spectratech.gr/en/product/28423/NVIS%20nVisor%20SX60
[2] http://www.instantreality.org/
[3] http://www.cyberglovesystems.com/products/cyberglove-II/overview

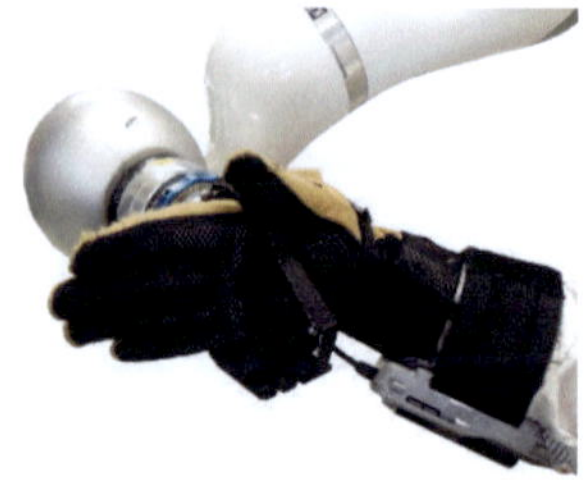

(a) Cyberglove

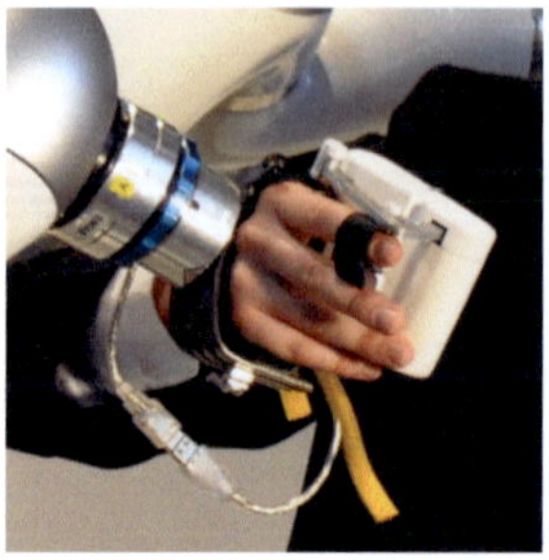

(b) Grasp Force Master

Figure 5.2.: Hand Interfaces used to control the robot hand: (a) the cyberglove allows teleoperated control of each finger of the DLR-HIT Hand II without force-feedback (b) the grasp-force master is a one degree of freedom device which enables force-feedback. It simplifies the command to "open" and "close" the end effector.

The glove tracks the motion of the human fingers, which are then mapped to the robotic hands with a point-to-point mapping. The DLR-HIT Hands execute the motion via a Cartesian impedance controller. Figure 5.2b shows a one DoF force feedback device (grasp force master, GFM) similar to the hand interface of the haptic device Sigma built by Force Dimension[4]. It allows the user to experience feedback of the grasping forces and simplifies the interaction between human and robotic hand. The GFM is used in the shared autonomy setup (and Study 2), where it replaces the teleoperated control of every single finger of the robotic hand.

To increase the level of autonomy of a telepresence system, not only planning algorithms but also object detection or scene analysis are needed. They provide information about which objects are in the remote environment and where they are located. Within the framework of this thesis, two methods have been integrated. First, a sequential analysis framework is utilized that is able to keep track of the changes to the remote scene over time [33]. Due to the expected lack of texture information in telepresence scenarios, the object recognition module of the framework is based on dense depth images, as produced by, e.g. stereo camera systems [86] or Kinect-like sensors. Specifically, a geometric matching approach [54], that was extended by a fast GPU based verification step, is used. Second, this scene analysis framework is used to initialize an algorithm that allows tracking of articulated objects in realtime [257]. In particular, both hands of the robot as well as the object to interact with can be tracked with depth-only

[4] http://www.forcedimension.com/products/sigma-7/overview

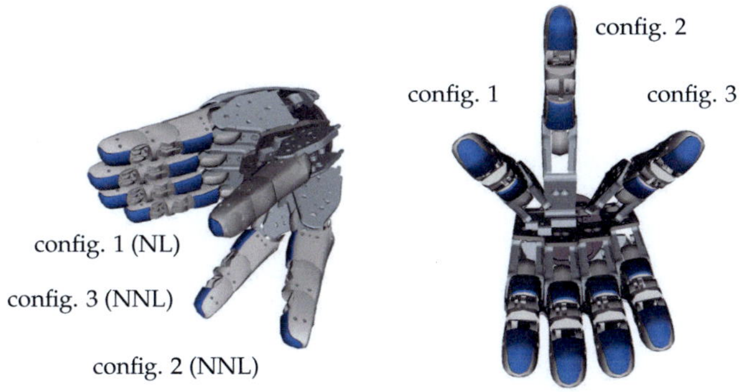

Figure 5.3.: Comparison of thumb configurations. Configuration 1 corresponds to the natural-looking hand (NL) used in both user studies. Configuration 2 and 3 are used in Study 1 and 2, respectively, as NNL hand.

information. This approach has the advantage that the relative pose between hands and object as well as the finger positions are computed and thus, no extrinsic calibration of the system is needed. Additionally, interpenetration of multiple interacting objects is avoided and contact detection by the robot is integrated to provide a physically correct estimate of the pose of the objects and fingers. The issue of estimate stability is handled by using many frames of data to converge over time to an accurate object pose estimate when the object is stationary, either relative to the camera (while on the table) or relative to one of the palms (while firmly grasped). The videos provided in the publications [243, 257] show the developed shared autonomy system.

5.2. Methodology: Study 1 and 2

The described shared autonomy approach is evaluated with two user studies. The studies have different foci according to the described challenges in grasping: Study 1 takes C2 and C3 into account; Study 2 uses the insights of Study 1 and focuses on visual assistance for C1. Both studies report the influence of the assistance functions on the objective and subjective performance of the system. We are particularly interested in the subjective workload of the human operator and his sense of presence. Since it is very likely that the kinematic configuration of the end effector has an effect on the performance of the system, each study uses two different hand dispositions: a natural-looking (NL) and a non-natural-looking (NNL) configuration. The configurations are physically realized using the DLR-HIT Hand II – a modular five-finger robotic hand [124], where all fingers have the same kinematics. The hand configuration is varied by changing the position of the thumb, as this finger has the largest influence on the grasping tasks (as it opposes all other fingers), and its position during the grasp execution is crucial to form a stable grasp. Also, this finger is the most visible one during the task execution. Figure 5.3 shows a comparison of the thumb configurations used in the user studies. Thumb configuration 1 denotes the NL hand used in both studies; it has a relatively small hand workspace. In Study 1, thumb configuration 2 is used as NNL hand. There, the thumb opposes the middle finger and has an opening angle of 60 °, which results in a large workspace. In Study 2, thumb configuration 3 is used as NNL hand, which ensures the same size of workspace as configuration 1. Both NNL hands are less intuitive to operate compared to the NL hand.

The NL hand has a thumb position that suggests kinematic equivalence to the human hand; however, the robotic hand has a highly restricted abduction/adduction angle of the thumb when compared to the human thumb. Nevertheless, we assume that the appearance induces the user to apply well-learned motor scripts with the NL hand, which might introduce difficulties in controlling the robotic hand when the subjective world model is incongruent with the perceived sensory data (i.e., the hand might not behave in the expected way). If system transparency is not high enough, this may lead to a trial-and-error strategy for grasping [39]. If the end effector is evidently morphologically different, as in the case of the NNL hands, this effect might not occur. Yet, the user then actively tries to learn the system, resulting in prolonged learning phases, i.e. higher cognitive load, decreased performance and a weaker sense of presence [219]. One possible solution to improve system transparency and to reduce the learning phase of the operator is the visualization of information about the system state. With this purpose in mind, we propose and evaluate the effectiveness of different visual assistance modes.

In summary, the main hypotheses for Study 1 and Study 2 are

H1 Performance increases when using assistance.

H2 Workload decreases when using assistance.

H3 The positive effects of assistance are higher for a NNL hand.

H4 A NL hand increases the level of presence and a NNL hand decreases it.

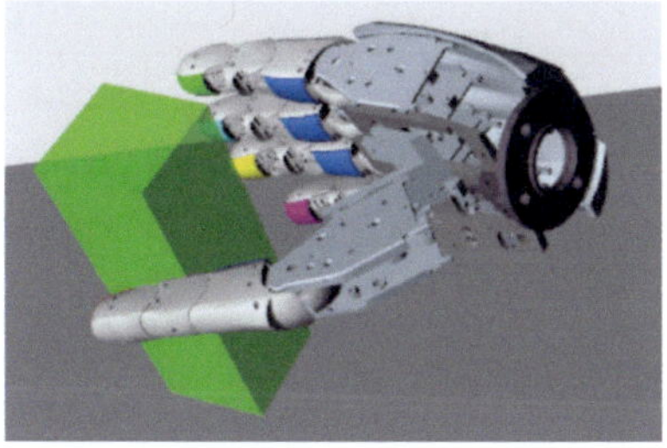

(a) Binary Assistance

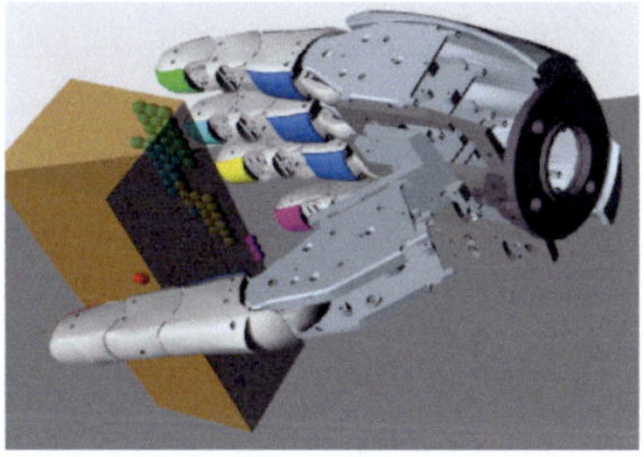

(b) Assistance with Regions

Figure 5.4.: Visual Assistance for Study 1.

5.2.1. Study 1 – Evaluation of Finger Positioning and Grasping

The first user study evaluates the developed assistance for finger adjustment and grasping (Challenges C2 and C3) in a real teleoperation system. A data-glove is used as hand interface, and two mapping strategies to control the robotic hand are implemented. The "assistance" and "no assistance" conditions are compared. Without assistance, the human controls the movement of every finger of the robotic hand via point-to-point mapping of the fingertips, and the computation of reachable independent contact regions (rICRs) does not influence the grasp at all. In the "assistance" condition, the online grasp planning computes rICRs (as described in Section 3.5). The result of the planning algorithm is presented to the human as visual assistance for the finger adjustment (Challenge C2) before execution of the grasp. In order to study the transparency of the system, two levels of visually displayed information for assistance are implemented. Both show if a force closure (FC) grasp is available for the current hand position relative to the object. In a simple binary assistance mode, a low amount of information is transmitted by changing the color of the object to green if a FC grasp is found and the object can be grasped (Figure 5.4a). For comparison, the other assistance mode displays the calculated contact regions on the object (Figure 5.4b). These regions give more information about the computed grasp, as the larger the region, the more robust the grasp is. Once the human decides to trigger the grasping action by closing his thumb and index finger, the robotic fingers are automatically directed towards a fingertip position inside their corresponding contact region (Challenge C3).

The experimental conditions for Study 1 are summarized in Table 5.2.

Table 5.2.: Overview of Experimental Conditions for Study 1

Kinematic Setup	Thumb Configuration 1 (NL)			Thumb Configuration 2 (NNL)		
Hand Control Mode	Direct Teleman.	Shared Autonomy		Direct Teleman.	Shared Autonomy	
Visual Assistance	None	Binary Assistance	Contact Regions Assistance	None	Binary Assistance	Contact Regions Assistance

5.2.2. Study 2 – Focus on Visual Assistance for Hand Positioning

The second user study focuses on assisting the hand positioning for grasping (Challenge C1). The study is conducted using the human machine interface HUG coupled to a virtual environment (not to the real robot), thus resulting in a high degree of experimental control as influences of the real robotic system are discarded. Again, in this study a visually assisted condition is compared to another one with no visual assistance. The visual assistance for hand positioning is based on the computation of a non-reachable FC grasp by iteratively allowing a new contact point for each finger outside of its reachable points, starting with the thumb. This non-reachable FC grasp is used to indicate the operator how to move the robotic hand such that the thumb has reachable points on the object that lead to a FC grasp. This movement is visualized with an arrow pointing from the fingertip of the thumb to its desired goal position on the object (Figure 5.5a). As soon as a stable grasp is found for a new hand pose, the arrow disappears. Then, the rICRs are displayed on the object surface (according to the results of Study 1), thus enabling assisted grasping. Instead of cybergloves, a one degree of freedom grasp force master (GFM) is used as interaction tool as its one DoF is sufficient to command the closing and opening of the robotic hand. The contact region for the thumb is colored red and the remaining regions for the opposing fingers are grey, to better distinguish the contact regions on both object surfaces.

In this study, the dynamic feasibility of the grasp is also evaluated. Given a kinematically feasible grasp, the maximum weight that the hand can lift is evaluated and compared to the current object's weight (as described in Section 3.6). The outcome is displayed with a billboard right next to the object, which continuously changes its color from red (object is not liftable with the current grasp) to green (object is liftable, chosen grasp is valid) as shown in Figure 5.5b. To

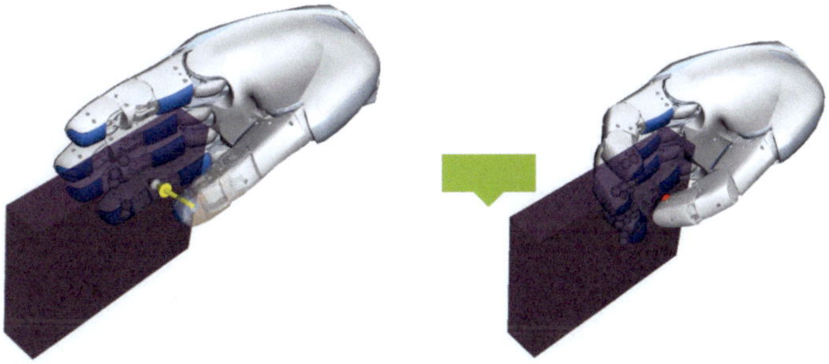

(a) Assistance for Hand Positioning (b) Assistance for Liftable Weight

Figure 5.5.: Visual Assistance for Study 2.

Table 5.3.: Overview of Experimental Conditions for Study 2

Kinematic Setup	Thumb Configuration 1 (NL)		Thumb Configuration 3 (NNL)	
Hand Control Mode	Shared Autonomy		Shared Autonomy	
Visual Info.	Contact Regions, Billboard for Weight		Contact Regions, Billboard for Weight	
Visual Assistance for C1	None	Assistance with Arrow	None	Assistance with Arrow

execute a grasp, the user simply has to close the GFM using the index finger, and the fingertips will be autonomously directed towards their corresponding contact region.

The experimental conditions for Study 2 are summarized in Table 5.3.

5.2.3. General Experimental Setup

Prior to the experiments, the users were informed about the objectives of the corresponding experiment and signed an informed consent. Then, subjects took seat in the human machine interface HUG and were informed about the experimental task and procedure, as well as the safety features of the interface (dead man pedal, magnetic clutch[5] at the handle). In both user studies, the participants used only the right arm of HUG to shorten the exercise phase prior

[5] Patent DE 10 2008 058 218.2

to the experiments. All the participants were right-handed, so this had no influence on the skill level of the operator. Their hand was attached to the right robot of HUG with the respective hand interface, and they could get familiar with the system being either coupled to a real robot (Study 1) or to a virtual environment (Study 2). The participants could always experience haptic feedback on their hand. Visual feedback is provided to the human using the HMD but the head tracking was not used. Instead, the viewpoint on the scenes was predefined and fixed to ensure comparability of the visual assistances between subjects. All participants had normal or corrected to normal vision, and were not paid to participate in the studies.

The studies are evaluated using objective performance data (for instance, time to complete the task and grasp quality) and subjective user feedback gathered through questionnaires applied after the experiment had finished.

5.3. Study 1 – Evaluation of Finger Positioning and the Actual Grasp

This section presents Study 1, where the effect of assistance for Challenges C2 and C3 is evaluated. The specific experimental setup is described first, followed by the numerical results and a discussion on their implications.

5.3.1. Experimental Setup

The NL and NNL hands used in this study are shown in Figure 5.6. The visual feedback is provided to the users in two windows, as shown in Figure 5.7a. The virtual scene where the assistance modes are visualized is rendered in 3D, and a smaller reference window shows the live stream of one SpaceJustin camera (in 2D). Both scenes provide the same viewpoint on the scenario. We found this dual feedback natural for the participants, in agreement with the results from other studies [43]. To reduce the influence of factors such as errors in object detection or forward kinematics, hand and object positions were optically tracked with a camera system (A.R.T.[6]); the optical markers can be seen in Figure 5.6 and Figure 5.7b.

[6] www.ar-tracking.com

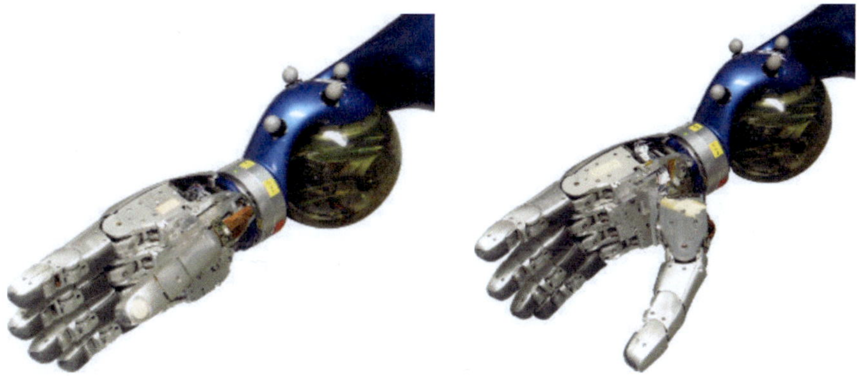

(a) NL Hand: Thumb Config. 1 (b) NNL Hand: Thumb Config. 2

Figure 5.6.: Two hand (thumb) configurations employed for user study 1.

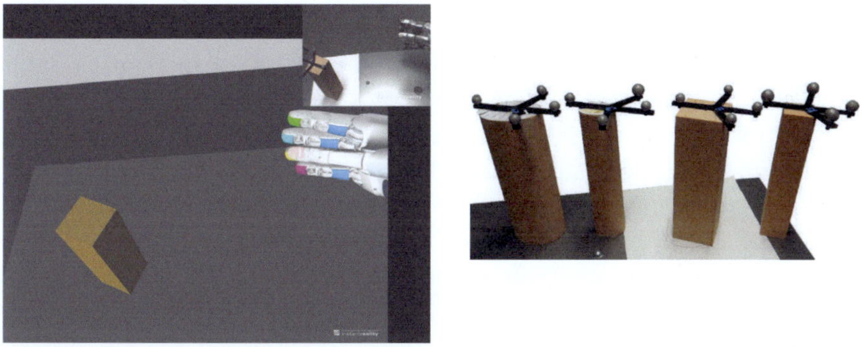

(a) View in Head-Mounted Display (b) Objects

Figure 5.7.: Experimental setup for study 1.

Procedure

The participants were instructed to grasp four objects with basic shapes, see Figure 5.7b: two cylinders and two cuboids with a small and a larger version of each. The heaviest object weighs 3.4 N. The small cylinder is easier to grasp with the NL hand and the large cylinder easier for the NNL hand due to the change in workspace related to the thumb configuration. The basic shapes were chosen to be able to study the influence of assistance with objects that are intuitive to grasp, therefore avoiding more complex objects that might lead to additional cognitive processing of the user.

Overall, three assistance conditions were tested: a no-assistance condition, and two conditions with assisted grasping. They differ in the amount of displayed information coming from the grasp planning process (Section 5.2.1). In all conditions, individuals were told to grasp the objects from the side with the fingertips of the robotic hand, to avoid object collisions and marker occlusions when performing the task. At the same time, they should perform secure grasps as quickly and reliably as possible. For each block of trials with 1 assistance mode and 1 thumb configuration, subjects completed a training trial first, trying to grasp an object that was easy to grasp with the current thumb configuration, and then they grasped the four objects one after the other. Altogether, a number of 6 blocks of grasp trials (2 thumb configurations x 3 assistance conditions) and 30 total grasp trials (6 blocks x 5 trials (1 training and 4 experimental trials)) had to be completed. After each grasp, the arm lifted the object 0.1 m straight up and rotated the object 30 ° towards the thumb and 25 ° forward to evaluate the grasp stability.

Sample

Twenty male participants were recruited from the student and staff population of the German Aerospace Center (Mean $M_{\text{AGE}} = 28.4\,\text{y}$; standard deviation $SD = 4.0\,\text{y}$) Five of the 20 participants had experience in using HUG but none in controlling the remote robot SpaceJustin. The other participants had no previous system experience. Overall, the subjects had a medium experience in using CAD programs (evaluated with the sentence "I work with CAD programs on a regular basis" on a 7-point Likert scale (1 "Does not apply", 7 "Fully applies") $M = 3.6, SD = 2.5$). Analysis of results revealed that previous experience with the system or with CAD programs had no influence on the performance in the study.

For the experiments, a within-subjects design with robotic hand (NL vs. NNL) and assistance (none vs. binary vs. regions) as within factors was utilized, i.e., each subject finished all six blocks of trials. The order of these blocks was systematically permuted to control for potential time effects like learning or fatigue. After each block, subjects filled out the NASA-TLX questionnaire (German version; [79]; given in Section A.1.1) and a specific questionnaire to evaluate interaction realism and usability (Section A.1.2 and A.1.3). Finally, participants compared and evaluated all assistance conditions (Section A.1.4)

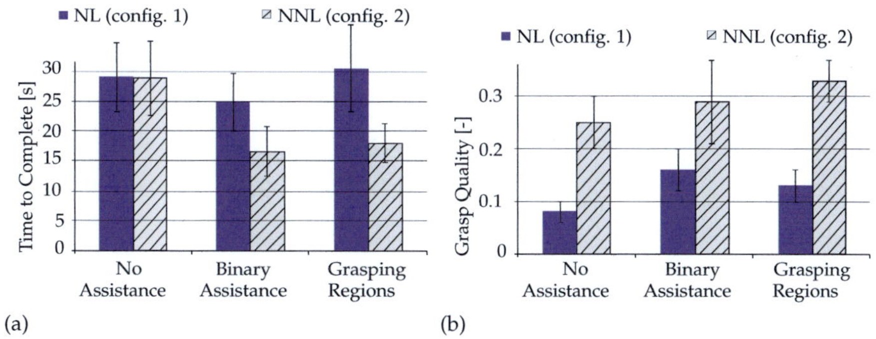

Figure 5.8.: Objective measures for Study 1; (a) time to complete the task (b) grasp quality

5.3.2. Results

The results are summarized in Table 5.4 and we start by giving the results of the objective data, namely time to complete (TTC), grasp quality, and task success.

Time to Complete (TTC, Figure 5.8a)
A repeated measures analysis of variance (ANOVA) with *hand* (NL vs. NNL) and *assistance* (none vs. binary vs. regions) as repeated measures was performed on the time to complete (TTC) each grasp. A highly significant main effect of *hand* ($F(1,17) = 10.9$; $p < 0.01$) occurred, revealing higher TTC when performing the task with the NL hand ($M = 28.1$ s) compared to the NNL hand ($M = 21.1$ s) when all the assistance modes are averaged. Moreover, a highly significant *assistance* main effect ($F(2,16) = 10.7$; $p = 0.001$) indicates significantly shorter TTC in the binary assistance condition ($M = 20.7$ s) compared to conditions without assistance ($M = 28.9$ s; $t(40) = 3.93$; $p < 0.001$). Yet, a significant two-way interaction effect *hand* x *assistance* ($F(2,16) = 4.9$; $p < 0.05$) revealed that the positive effect of assistance conditions is stronger when using the NNL hand. In this case, TTCs were significantly shorter when having binary ($M = 16.6$ s) or grasp region assistance ($M = 18$ s) compared to the conditions without assistance ($M = 28.8$ s; both $ts(18) > 3.7$ and $ps < 0.01$). No significant (n.s.) difference between both assistance conditions was found ($t(18) = 0.7$; n.s.).

Grasp quality (*100, Figure 5.8b)
Repeated measures ANOVA performed on the grasp quality revealed a highly significant *hand* main effect ($F(1,17) = 45.2$; $p < 0.001$), with better results for

Table 5.4.: Overview of results of Study 1 with means (standard deviation)

	Anthropomorphic Hand			Non-anthropomorphic Hand		
	No Assistance	Binary	Regions	No Assistance	Binary	Regions
TTC, Time To Complete [s]	29.0 (11)	24.8 (8.9)	30.5 (14.8)	28.8 (12.6)	16.6 (8.2)	18.0 (6.4)
Grasp Quality [-]	.08 (.05)	.16 (.08)	.13 (.07)	.25 (.01)	.29 (.16)	.33 (.09)
Workload (very low: 0; very high: 20)	8.3 (3.6)	6.5 (3.2)	6.9 (3.0)	8.4 (3.7)	5.7 (3.0)	5.6 (3.2)
Usability (none: 0; very high: 36)	-	25.8 (5.9)	23.6 (7.2)	-	27.0 (4.4)	26.4 (5.4)
Presence (none: 0; very high: 24))		12.9 (4.4)			13.9 (4.5)	
Disturbance (none: 1; very high: 7)		4.5 (1.8)			2.4 (1.0)	
Habituation (none: 1; very high: 7)		4.3 (1.7)			5.5 (1.2)	

the NNL hand ($M = 0.29$) than for the NL hand ($M = 0.12$). Additionally, a significant *assistance* main effect ($F(2, 16) = 10;\ p < 0.01$) indicated worst grasping quality without assistance ($M = 0.16$), better quality with binary assistance ($M = 0.23$) and grasp regions assistance ($M = 0.23$). Again, a significant two-way interaction effect *hand* x *assistance* occurred ($F(2, 16) = 4.1;\ p < 0.05$), i.e. grasp quality was significantly better when working with the contact regions and NNL hand compared to the NL hand.

Task Success

The automated test routine for grasp stability revealed no significant effects (all $Fs < 2.2$), sometimes objects fell out of the hand or changed the pose inside the hand in all the tested assistance modes. This happened three to five times out of 160 grasps per assistance mode. For instance, with the grasp regions assistance all grasps were stable when using the NL hand, while only 92 % were stable with the NNL hand.

Subjective Data

Besides the objective measures, participants answered questions regarding their workload. After finishing all blocks with one hand variant, the sense of presence and the interaction with that robotic hand was evaluated. Also, the evaluation of the assistance functions is presented.

Workload

A repeated measures ANOVA on the NASA-TLX overall score (range from 0 to 20) revealed no *hand* condition main effect ($F(1, 20) = 1.8$; n.s.), but a highly significant *assistance* condition main effect ($F(2, 19) = 14.3;\ p < 0.001$). For both hand configurations, we found significantly lower workload scores for the binary feedback conditions ($M_{\text{NL}} = 6.4;\ SD_{\text{NL}} = 3.2;\ M_{\text{NNL}} = 5.7;\ SD_{\text{NNL}} = 3.0$) and the grasp region conditions ($M_{\text{NL}} = 6.9;\ SD_{\text{NL}} = 3.0;\ M_{\text{NNL}} = 5.6;\ SD_{\text{NNL}} = 3.2$) compared to the no assistance conditions ($M_{\text{NL}} = 8.3;\ SD_{\text{NL}} = 3.6;\ M_{\text{NNL}} = 8.4;\ SD_{\text{NNL}} = 3.7$; all $ts(20) > 2.2$; $ps < 0.05$). This effect was particularly evident when working with the NNL hand (both $ts(20) > 4.7$; $ps < 0.001$). The average workload scores in the binary feedback and grasp regions condition did not differ significantly (both $ts(20) < 0.9$).

Sense of Presence

The sense of presence was evaluated with a four-item-questionnaire that evaluated in a range from 0 to 24 the consistency of expectation, natural interaction, effort, and tool embodiment (the sensed fusion between your own and the robotic hand). There were no significant main effects of the configuration of the hand on the sense of presence ($t(19) < 0.9$, n.s.).

Interaction with robotic hands
We asked participants to which degree they felt disturbed when working with the respective thumb positions of the robotic hand "The thumb position severely disturbed me during interaction" (does not apply: 1; fully applies: 7). Subjects reported that they felt significantly more disturbed by the thumb position of the NL hand ($M = 4.5$; $SD = 1.8$) compared to the NNL hand ($M = 2.4$; $SD = 1.0$; $t(19) = 5.2$; $p < 0.001$). Moreover, participants also indicated that they got used to the thumb position of the NNL hand after a short phase of training; for the item "After a short period of training, I did not longer worry about the thumb position" ratings were significantly higher for the NNL hand ($M = 5.5$; $SD = 1.2$) compared to the NL hand ($M = 4.3$; $SD = 1.7$).

Assistance functions
Participants had to answer the following question (all scales using the 7-point Likert scale): "Even before completing the grasp I was sure that I would securely grasp the object". When working without assistance function, significantly lower ratings were reported ($M = 2.3$; $SD = 1.0$) compared to binary ($M = 5.0$; $SD = 1.4$) and grasp regions assistance ($M = 5.8$; $SD = 1.1$; both $ts(19) > 6.4$; $ps < 0.001$). Moreover, ratings in the grasp regions conditions were significantly higher than in the binary feedback condition ($t(19) = 2.2$; $p < 0.05$). The users perceived that binary assistance did not give enough information about possible grasps; the item was "When working with binary feedback there were situations in which I did not know how to modify the hand position to find an optimal grasp" ($M = 4.0$; $SD = 1.8$).

5.3.3. Discussion

Both objective measures revealed better grasping performance with assistances for both hand configurations, and a decrease in workload was reported (hypotheses H1 and H2 confirmed). Moreover, the assistance effect was stronger for the NNL hand, as hypothesized (H3 confirmed). Yet, the assistance showing the grasp regions leads to a slightly higher TTC than in the case of binary feedback. As the grasp regions give a more detailed description of the grasp robustness, the higher TTC results in grasps with higher quality for the NNL hand, where human intuition is not so helpful as the configuration is less anthropomorphic. Overall, the NNL hand leads to a reduction of TTC and to an increment in grasp quality compared to the NL hand.

There was no significant difference in objective performance measures between the two assistance modes, although people felt more secure when being assisted with the contact regions. Comparing the hand configurations, we noticed that the grasp quality decreases and the TTC increases for the NL hand when using the assistance with regions (compared to the binary assistance). This can be induced by the differences in kinematic and dynamic capabilities between the human and the NL hand. In general, it is important to place the thumb in opposition to the other fingers for getting a robust grasp. The human can simply move the thumb to ensure this opposition. Although the NL hand looks anthropomorphic, the restricting limit on the abduction/adduction angle and the lack of rotation for the thumb impedes such behavior. For a proper hand positioning, an additional hand rotation is needed. This fact was empirically verified when comparing the overall rotational movement of the hands; the users rotated more the NL than the NNL hand. Additionally, one could observe that the participants rotated the hand the most when using binary assistance, as they had no information on "how close" they were to a valid grasp. The display of the rICRs created the feeling of "being close to a good grasp," and people usually relied more on translational exploration (rather than rotational exploration) to improve the grasp, which resulted in a higher TTC and often in less stable grasps. The contact regions improve system transparency and trust, as indicated by the participants, while the additional visual information did not result in higher workload compared to a simpler on/off visualization.

There was no significant effect regarding the sense of presence comparing the hand variants (H4 not confirmed). Moreover, the sense of presence was even rated slightly higher for the NNL hand compared to the NL hand. As the NL hand looks more human like, we expected people to handle that hand more intuitively, which was not confirmed in the subjective data. In fact, participants felt more disturbed by the thumb configuration in the NL hand, although there was no difference in TTC without assistance between both hand configurations. Reactions on the thumb positions included: Thumb configuration 1 (NL) is "more intuitive but with smaller workspace", and "the opening angle is too small". Configuration 2 (NNL) is "less intuitive to control", "the thumb has an unnatural position, but is easier to handle after some training", "small objects are hard to grasp", and "the larger opening angle of the thumb is helpful".

With no assistance users sometimes grasped with a poorly positioned thumb, which lead to unstable grasps We additionally observed that poor fingertip grasps sometimes moved the object within the hand producing very stable power grasps (although the intention was to make a precision grasp). In the cases where assisted grasping failed the test, it was mainly due to two reasons: tracking errors and torque limits in the thumb. Due to the modular hand design, the thumb has the same motors as the other four fingers. As the calculation of rICRs does not consider the dynamics of the fingers, the thumb sometimes could not exert enough forces to counteract the object's weight, and the object was dropped during the test routine. This happened especially for the NNL hand as the thumb here has to travel more space to reach the object surface, thus resulting in configurations of the finger that only allowed very low forces applied on the object.

5.4. Study 2 – Focus on Visual Assistance

This section presents Study 2, where the effect of assistance for hand positioning relative to the object (C1) is evaluated. As the shared autonomy approach was found to increase system performance in a real teleoperation scenario in Study 1, this study focuses on the visual assistance under the influence of different kinematics of the end effector. The visual assistance is evaluated in a virtual reality (VR) scenario. The experimental setup is first described, followed by the summary of results and a discussion on the findings.

5.4.1. Experimental Setup

For this study, the thumb of the NNL hand is not opposing the middlefinger, but it was moved even further such that the hand looks like a left hand, as shown in Figure 5.9. Therefore, the user feels like controlling a left hand with his right hand in the experiment. Both haptic and visual feedback are provided to the users during the study. They could see a virtual scene through the head-mounted display, and perceived force feedback on their hand when a collision happened in the VR environment. The scene includes a static cuboid placed on a surface, and the virtual hand. The cuboid as well as the thumb were displayed in a semi-transparent way to avoid complete occlusions of the fingers or visual aids.

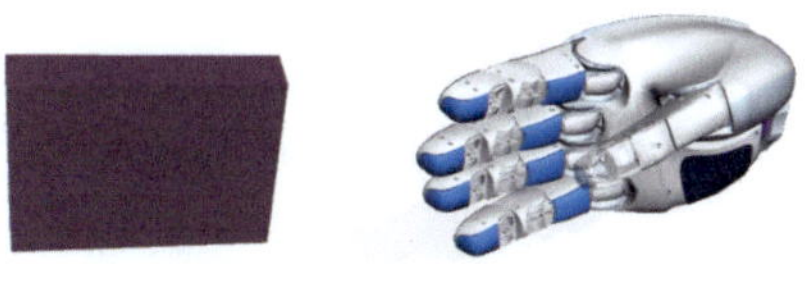

(a) NL Hand: Configuration Right

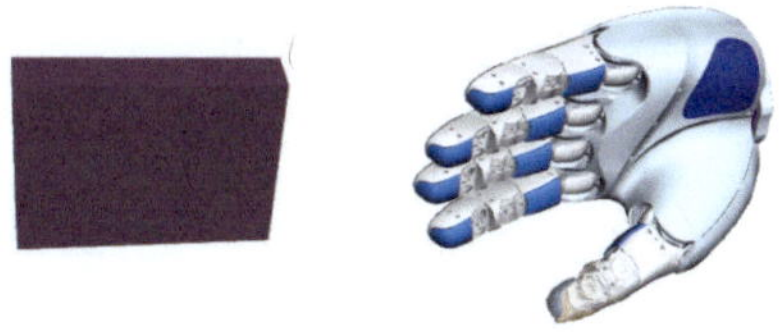

(b) NNL Hand: Configuration Left

Figure 5.9.: Two configurations of the DLR-HIT Hand II for Study 2.

Procedure

Participants had to grasp the cuboid from the right side using the fingertips, without actually lifting the object. Specifically, the hand position and orientation had to be optimized until reaching a hand pose where the finger joint configuration allowed exerting a force high enough to counteract the weight of the object. The difficulty of the reaching and grasping task is modified by using different relative initial positions between the cuboid and the hand (Figure 5.10) to demand various reaching strategies. A larger hand rotational motion was required for pose 1 than for pose 3, while for pose 2 mainly a translational movement was required. The different object poses were also chosen to avoid that subjects simply memorized the optimal grasping strategy when repeatedly performing the same grasp and it was taken care that the visual aids used in the study were not obstructed. As soon as a valid grasp for the current hand pose was found, contact regions were displayed on the cuboid, showing the final positions of the fingers on the cuboid surface if the hand would be closed (Figure 5.5). Participants could close the GFM whenever they felt the hand position was appropriate, and the scene would display the hand closed around the object (Figure 5.5b). The task was only completed when the grasp was valid, i.e., the billboard visualizing the liftable weight of the object was green.

In all conditions subjects were given a time limit of 60 s for the task, so they should try to grasp the cuboid as quickly as possible with a valid grasp. The task was finished and data logging was stopped when the user pressed the grasping mechanism of the GFM and afterward released the dead man pedal of HUG. A within-subjects design was utilized, with hand variant (NL vs. NNL), hand orientation aid (arrow on vs. off) and object pose as within-subject factors. The order of the within factors was counterbalanced to control for potential time effects (like learning or fatigue). In each experimental block, subjects could train the grasps once (with one random object pose), before starting with the

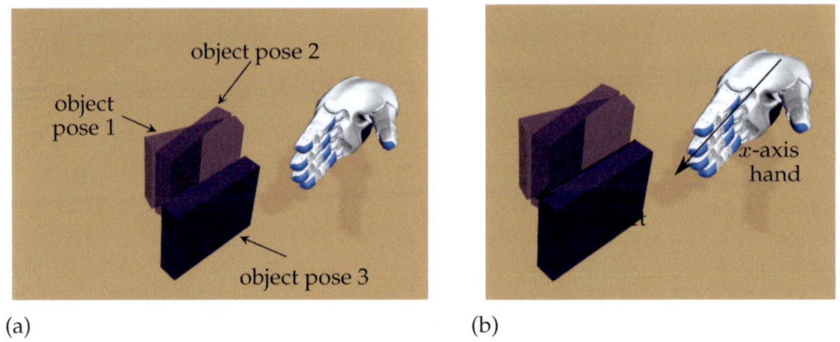

Figure 5.10.: Positions of the Cuboid for Study 2. The angle between the x-axis of the object and the x-axis of the initial pose of the hand are: object pose 1: 28 °, pose 2: 5 ° and pose 3: 11.5 °

actual grasps of the object in the three predefined poses. Altogether, a number of 4 (1 training grasp + 3 experimental grasps) x 2 (thumb configurations) x 2 (hand orientation aid) = 16 trials had to be completed.

After each experimental block, the NASA-TLX questionnaire was administered for measuring mental workload. The sense of presence was indicated by answering the item "How strong was your sense of presence in the virtual environment (no sense of presence: 0 % to sense of presence like in the real world: 100 %)". In Study 1, participants filled out the German version of the "Presence Questionnaire" (PQ, [186]) after all blocks of trials for one hand, which lead to no significant results. In this study, the questionnaire was administered after each experimental block (within one hand variant) with the subscales "spatial presence", "interface quality" and "involvement and motivation". Also, the usability of the assistance was evaluated with the "AttrakDiff" questionnaire with the subscale "quality" [80]. Furthermore, the German version of the "Immersive Tendencies Questionnaire" with the subscales "emotional involvement" was administered after having finished the whole experiment [186]. All questionnaires are given in Section A.2.

Secondary Task

During all of these trials, a secondary task had to be performed, to measure attention allocation toward the virtual scene vs. the real environment. With this task, we validate the values given by the participants in the PQ questionnaires. Specifically, subjects had to respond as quickly as possible to a sound (0.5 s long buzzer-like sound similar to the sound of a cell phone vibration alert) by pressing a foot pedal. The sound was played randomly every six to nine seconds.

Sample

Sixteen (one female, fifteen male) subjects from the student and staff population of the DLR with an average age $M = 29.9\,\text{y}$ ($SD = 5.8$) participated in the study. Seven subjects had never worked with HUG, five had been working with it during one or two sessions, and the remaining four were expert users (> 20 sessions). Participants' sensorimotor skills were assessed with the sensorimotor coordination (SMC) module of the Vienna Test System[7] prior to the main experiment. Although both abilities were distributed normally (Kolmogorov-Smirnov-Zs = 0.69; non-significant, n.s.), the mean percentile rank of $M_{\text{SMC}} = 72$ ($SD = 24$) indicates above-average abilities of the recruited sample.

5.4.2. Results

The results are summarized in Table 5.5 and first, the results for the objective measures are given. We considered time to complete the task (TTC), time to orient the hand (TOH), the rotational movements, the task success, as well as the amount of liftable weight.

Time to Complete the task (TTC)

TTC is the time interval between the first movement of the hand until the final, successful grasp. A repeated measures analysis of variance (ANOVA) with hand variant (NL vs. NNL), hand orientation aid (on vs. off) and object pose (1 vs. 2 vs. 3) as repeated measures was performed on the TTC measure. While there was no significant main effect of hand orientation aid ($F(1, 14) = 0.92$; n.s.) and object pose ($F(2, 13) = 1.13$; n.s.), a marginally significant hand variant main effect ($F(1, 14) = 2.31$; $p < .10$, $\eta^2 = .14$, one-tailed testing) occurred: as expected, TTC with the NNL hand was longer ($M = 28.3\,\text{s}$; $SD = 17.4$) than with the NL hand ($M = 23.5\,\text{s}$; $SD = 13.4$).

Time to Orient Hand (TOH, Figure 5.11a)

For a more detailed analysis of grasping procedure, the time interval between the first hand movement and a successful alignment of hand pose such that contact regions could be displayed, was analyzed. ANOVA indicated a significant main effect of the hand orientation aid ($F(1, 14) = 4.29$; $p < .05$, $\eta^2 = .25$,

[7] http://www.schuhfried.com/viennatestsystem10/vienna-test-system-vts/

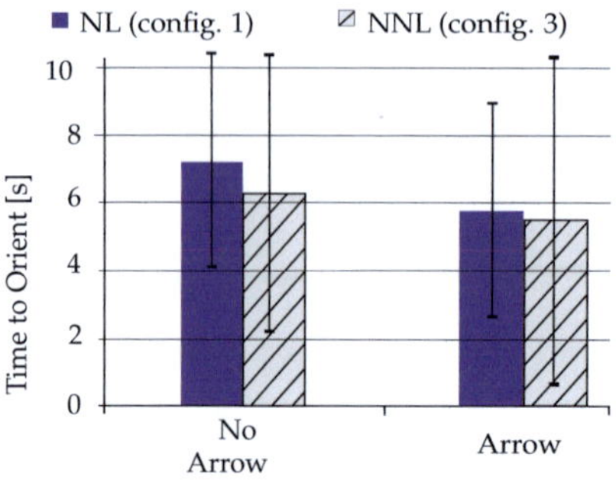

(a)

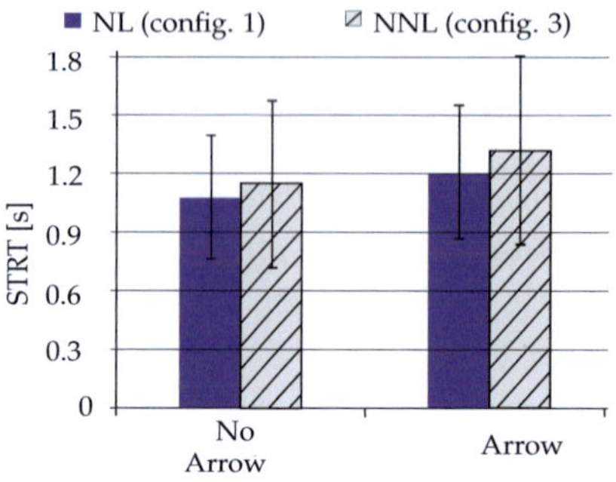

(b)

Figure 5.11.: Objective measures for Study 2; (a) time to orient the hand (b) secondary task reaction time

one-tailed testing), with shorter TOH when the aid was activated ($M = 5.7$ s; $SD = 3.0$) compared to conditions without aid ($M = 6.7$ s; $SD = 3.5$). No such effects were evident for hand variant and object pose (both $Fs < 2.24$).

Rotational Movements

Furthermore, we explored the cumulative rotational movements (in degrees) during the whole grasping movement. Indeed, ANOVA showed that with hand orientation aid less rotations were performed ($M = 293.7\,°$; $SD = 143.1$) than without aid ($M = 376.3\,°$; $SD = 170.9$), reaching the conventional level of significance ($F(1, 14) = 4.29$; $p < .05$, $\eta^2 = .27$). No significant effect occurred for hand variant. Yet, as expected, the object pose had a significant effect ($F(1, 14) = 4.78$; $p < .05$, $\eta^2 = .27$). Pairwise t-tests with Bonferroni correction indicated that the higher rotation values were obtained with object pose 1 compared to pose 2 and pose 3 (pose 1 vs. $3 < .05$; pose 1 vs. $2 < .10$).

Task Success

Analyzing the task success (i.e. a successfully performed grasp within the 60 s time limit) revealed a similar result pattern, with a significant hand variant main effect ($F(1, 13) = 5.54$; $p < .05$, $\eta^2 = .30$) and no significant main or interaction effects of the other factors (all $Fs < 1.1$; n.s.). Using the NNL hand decreased performance with an average task success of $M = 79\%$ ($SD = 16\%$) compared to the NL hand with $M = 89\%$ ($SD = 25\%$).

Liftable Weight

The mean liftable weight during the grasp procedure was analyzed next. Although the task specification did not include maximizing this weight, participants needed to reach a minimum threshold to successfully end the experimental trial. No significant hand variant effect occurred ($F(1,13) = 0.46$; n.s.), but a marginally significant hand orientation aid effect ($F(1,13) = 3.5$; $p = .08$, $\eta^2 = .21$) and object pose effect ($F(2,12) = 3.6$; $p = .06$, $\eta^2 = .38$). The mean liftable weight was higher with the arrow activated ($M = 0.28$, $SD = 0.10$) compared to no arrow assistance ($M = 0.25$, $SD = 0.10$). Moreover, mean weight for object pose 3 was significantly higher than for object pose 1 as revealed by subsequent t-tests ($M_{\text{pose1}} = 0.23$; $M_{\text{pose2}} = 0.27$; $M_{\text{pose3}} = 0.29$; $p_{1-3} < .05$); all other contrasts did not reach significance.

Secondary Task Reaction Time (STRT, Figure 5.11b)
Performing an ANOVA on the reaction times on the sound signal showed a marginally significant main effect of the hand variant ($F(1,13) = 2.39$; $p < .10$, $\eta^2 = .16$, one-tailed testing), i.e. STRTs were longer when grasping with the NNL ($M = 1.23\,\text{s}$; $SD = 0.35$) compared to the NL hand ($M = 1.15\,\text{s}$; $SD = 0.30$). Moreover, a significant main effect of hand orientation aid occurred ($F(1,13) = 3.32$; $p < .05$, $\eta^2 = .20$, one-tailed testing), with longer STRTs when the hand orientation aid was activated ($M = 1.27\,\text{s}$; $SD = 0.35$ vs. $M = 1.11\,\text{s}$; $SD = 0.35$). No significant main effect for the object pose was evident ($F(2,12) = 2.76$; n.s.).

Interestingly, explorative analyses revealed that the immersive tendencies predict the average STRT, i.e. the answers to questionnaires correlate to the measured reaction times. This is indicated by a multiple regression analysis with "emotional involvement" ($\beta = .67$, $p < .01$) and "focus" ($\beta = .33$, $p < .10$) as predictors (adjusted $R2 = .60$, i.e. 60 % of the variance).

Subjective Measures

Mental Workload
Analyzing the weighted overall NASA-TLX sum score (scale ranging from 0 to 20) revealed no significant main effects. While no main effect was found for the hand variant ($F(1,14) = .64$; n.s.), a marginally significant effect of the hand orientation aid occurred ($F(1,14) = 3.63$; $p < .10$, $\eta^2 = .21$). i.e. workload decreased when using the aid ($M_{\text{NoAid}} = 9.7$; $SD = 3.5$; $M_{\text{Aid}} = 8.4$; $SD = 3.7$). However, the isolated analysis of the NASA-TLX mental demand item ("How mentally demanding was the task?") showed that with the NNL hand ratings tended to be higher ($M = 9.9$; $SD = 6.7$) compared to the NL hand ($M = 8.4$; $SD = 5.4$; $F(1,14) = 3.6$; $p < .10$, $\eta^2 = .20$).

Table 5.5.: Overview of results of Study 2 with means (standard deviation)

	NL Hand		NNL Hand	
	No Arrow	Arrow	No Arrow	Arrow
TTC, Time to Complete [s]	25.1 (15.5)	22.0 (4.0)	29.5 (4.5)	27.0 (5.4)
TOH, Time to Orientate Hand [s]	7.2 (3.2)	5.8 (3.2)	6.3 (4.1)	5.5 (4.8)
Rotational Movement [°]	363.3 (188.1)	261.1 (114)	389.2 (212.9)	326.2 (227.9)
Task Success [%]	86.1 (20.7)	91.7 (20.7)	75.9 (30.2)	82.4 (26.2)
Liftable Weight [N]	0.24 (0.10)	0.27 (0.10)	0.26 (0.12)	0.28 (0.12)
STRT, Secondary Task Reaction Time [s]	1.08 (0.32)	1.21 (0.34)	1.15 (0.43)	1.32 (0.48)
Workload (very low: 0; very high: 20)	9.0 (3.4)	8.4 (3.6)	10.3 (4.3)	8.3 (4.8)
Sense of Presence [%]	64.4 (25.0)	62.9 (23.0)	53.3 (23.0)	60.8 (24.6)
Usability (none: 0; very high: 36)	28.7 (6.2)	31.3 (7.5)	28.2 (5.2)	29.0 (8.1)

Next, we analyzed the user's experience with the haptic user interface (novice vs. moderate experience vs. expert) in an ANOVA performed on the sum score with user experience as additional between-subjects factor. First, there was a marginally significant main effect of user experience on mental workload ($F(2,6) = 4.53$; $p < .10$, $\eta^2 = .69$), showing that the more experience subjects had, the lower the reported workload ($M_{\text{Novices}} = 9.3$; $SD = 3.4$; $M_{\text{ModExp}} = 8.1$; $SD = 7.6$; $M_{\text{Experts}} = 5.4$; $SD = 4.4$). Furthermore, a significant interaction effect hand orientation aid x user experience was evident ($F(2,4) = 7.28$; $p < .05$, $\eta^2 = .79$). This effect means that the magnitude of the positive effect of the hand orientation aid is higher the less experienced the subject is (novices: $M_{\text{NoAid}} = 10.9$; $M_{\text{Aid}} = 8.0$; Moderate Experience: $M_{\text{NoAid}} = 9.0$; $M_{\text{Aid}} = 7.3$; experts: $M_{\text{NoAid}} = 5.4$; $M_{\text{Aid}} = 4.7$). There were no effects of user experience on the other measures.

Sense of Presence (SP)
Subjects had to indicate how much they felt that they were physically present in the virtual environment (from 0-100%). Here, ANOVA indicated 1) a significant hand variant main effect ($F(1,14) = 5,15$; $p < .05$, $\eta^2 = .27$) and 2) a significant hand variant x hand orientation aid interaction effect ($F(1,14) = 5.18$; $p < .05$, $\eta^2 = .27$). First, SP was higher for the NL ($M = 63.7\%$; $SD = 23.4$) compared to the NNL hand ($M = 57\%$; $SD = 22.6$). Second, a marginally significant positive effect of the hand orientation aid was evident for the NNL hand variant ($M_{\text{NoAid}} = 53.3\%$; $SD = 23$; $M_{\text{Aid}} = 60.8\%$; $SD = 24.6$; $p_{\text{diff}} < .10$, one-tailed t-test), while no significant difference was found for the NL hand ($M_{\text{NoAid}} = 64.4\%$; $SD = 25$; $M_{\text{Aid}} = 62.9\%$; $SD = 23$).

The moderating impact of sensorimotor skills (SMC high vs. low scorers) was explored. A highly significant interaction effect of hand orientation aid and SMC ($F(1,9) = 23.15$; $p = .001$, $\eta^2 = .72$) occurred. For the low SMC subjects the hand orientation aid did not improve the sense of presence ($M_{\text{NoAid}} = 55.4\%$; $SD = 36.2$; $M_{\text{Aid}} = 52.8\%$; $SD = 34.0$), while for subjects with high SMC scores, the hand orientation aid significantly enhanced the SP ($M_{\text{NoAid}} = 61.0\%$; $SD = 35.0$; $M_{\text{Aid}} = 70.4\%$; $SD = 33.0$). In addition to the SP-item reported above, the subscales from the presence questionnaire were explored. Yet, ANOVA with the hand variant as within factor did not reveal any significant effects. Still, explorative correlation analyses indicated significant positive relationships between the average SP as measured with the one-item approach and the subscales "interface quality" ($r = .65$; $p < .01$) and "involvement and motivation" ($r = .65$; $p < .01$). The subscale "spatial presence" did not reach significance ($r = .22$; n.s.).

Usability
ANOVA did not reveal any significant effects (all $Fs < 1.5$). As expected, the highest usability rating was reported for the NL hand with hand orientation aid activated ($M = 31.3$; $SD = 7.5$).

Interaction with robotic hands
Additionally, after the whole experiment the subjects rated whether they felt disturbed by the hand configuration during interaction (1: does not apply at all, 7: fully applies). Indeed, a significant effect occurred in a t-test when comparing both hands, with lower values for the NL hand ($M_{\text{NL}} = 2.1$; $SD = 1.1$; $M_{\text{NNL}} = 3.4$; $SD = 1.9$; $t = 2.5$; $p < .05$). Finally, subject ratings for the item "After a short period of training, I did not longer worry about the thumb

position" revealed non-significant higher values for the NL compared to the NNL ($M_{\mathrm{NL}} = 5.9$; $SD = 1.5$; $M_{\mathrm{NNL}} = 5.4$; $SD = 1.7$; $t = 1.5$; $p = .15$).

5.4.3. Discussion

The objective performance measures "TTC", "TOH" and "liftable weight" directly measure the behavior during the two phases of grasping. The approach phase is mainly influenced by the hand orientation aid, as no grasp (and hence no rICRs) are found and displayed. Once there is a hand configuration with a grasp, the hand orientation aid is not necessary anymore and the goal is finding a grasp able to counteract the weight of the object. Thus, the objective performance measure mainly influenced by the hand orientation aid is the TOH. The hand positioning aid served its function well, as reflected by the significantly reduced hand orientation times for positioning the hand with respect to the object, and the reduced overall rotational movements (Hypothesis H1 confirmed). TTC shows no significant influence by the hand orientation aid, but interestingly, the liftable weight is higher with assistance (even though the orientation aid does not include information about grasp robustness). Workload was decreased using the assistance for hand orientation (H2 confirmed).

Using the objective measures, hypothesis H3 (assistance effects are higher for a NNL hand) cannot be confirmed. Nevertheless, the overall task performance using the NNL hand was worse compared to the NL hand (completion times were longer and the number of successful grasps was lower). This can be explained by the obvious mismatch between the robotic and the user's hand, which triggers a familiarization or learning process that results in a new internal representation of the input-output mapping. Yet, the results of the mean liftable weight were not significantly different for the hand variants. This finding might be explained by the fact that subjects paid greater attention to the visual assistance for hand fine adjustment (contact regions and billboard) when working with the NNL hand once they learned the mapping, which would support hypothesis H3.

Consistently, when grasping with the NNL hand more attention was attributed to the virtual scene (as indicated by longer STRTs) due to the learning process described above and a stronger focus on visual assistance. Additionally, the strong link between individuals' immersive tendencies and the STRT also suggests that individuals were emotionally more involved in the virtual task. The higher involvement did not result in a stronger sense of presence when working with the NNL hand, but in a higher mental workload. Seemingly, the interface or mapping to the NNL hand was more difficult compared to the NL hand which led to a weaker sense of presence despite the high level of task involvement (H4 confirmed).

A stronger attention focus on the virtual scene was not only triggered by the NNL hand but also by the hand orientation aid, as additional information had to be processed. Yet, the assistance function obviously facilitated the spatial interpretation of the virtual scene, decreasing subjective workload - particularly for less experienced operators. The improved spatial awareness and system transparency also led to improved presence ratings for the NNL hand. For both hands, sense of presence was facilitated for individuals with high sensorimotor skills. Apparently, these subjects were better able to interpret and intuitively understand the hand orientation aid and thus experienced a higher level of interactivity and presence in the virtual scene.

5.5. The DLR Assistive Robotics System

A setup for assistive robotics is used to evaluate the assistance for hand positioning with virtual fixtures. The focus of this robotic system lies in controlling a robotic manipulator with a bio-signal based HMI in order to assist people with disabilities. In particular, the manipulator is a DLR light-weight robot arm III [88] equipped with the DLR-HIT Hand II, the five-finger robotic hand also used in the previous studies [124]. This version of the hand uses the commercially available thumb configuration (NL hand, thumb configuration 1, shown in Figure 5.3). The HMI is realized with two interfaces: one uses neural signals [89], while the other detects remaining muscular activity with a surface EMG [213]. With both interfaces, the continuously measured signals are decoded in three Cartesian (translational) degrees of freedom and a discrete grasp trigger, as shown in Figure 5.12. These Cartesian degrees of freedom are mapped to the velocity of the end effector position of the manipulator [214]. Thus, during teleoperation the user has no direct influence over the rotational behavior of the robotic arm which limits its workspace and the approach directions to objects. Additionally, the performed grasp by the fingers is preprogrammed and executed upon the grasp trigger [89]. The operator shares the same workspace as the robotic manipulator and therefore, gets natural visual and auditory feedback.

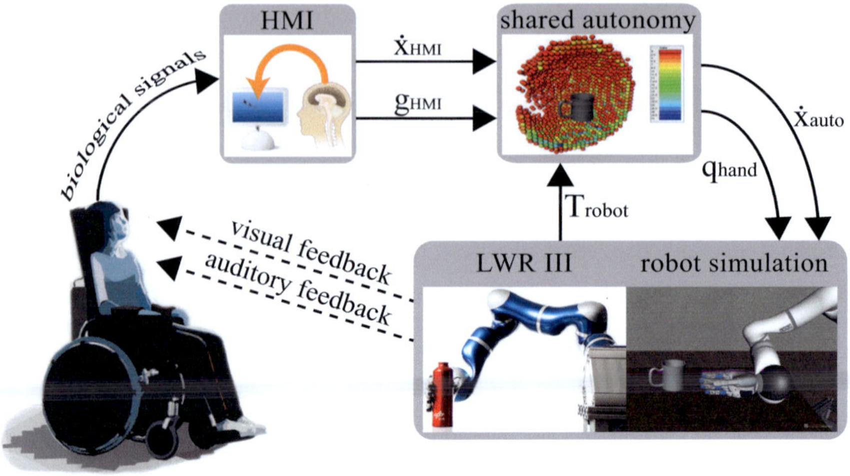

Figure 5.12.: Semi-autonomous grasping with the DLR Assistive Robotics System with courtesy of Jörn Vogel.

Although very promising to increase self-determination of disabled people, in addition to the workspace restriction mentioned above, the velocity directions obtained from bio-based signals are noisy which requires the user to learn how to control the arm during direct teleoperation in order to approach objects. Slightly inaccurate commands can lead to grasp failures which make grasping and manipulation of objects a slow and tedious task. Therefore, the level of autonomy of this system was increased to incorporate assistance for hand positioning, which is based on the methods described in Chapter 4. The operator remains in full control of the translational velocity of the end effector. These velocity commands are enhanced using virtual fixtures. They are not preprogrammed to support one approach direction to the object, but they are adapted online according to the movement suggested by the human. This is realized with the help of the graspability map for each object, which covers all hand poses around the object and rates them according to a quality criterion. During runtime, the commanded velocities and the current position of the robot with respect to the available objects are used to determine the object to be grasped and to search for a stable grasp pose in the database. To simplify the motion of the robot towards the object, the user-commanded velocity is gradually modified to converge to the selected grasp pose. The modification of the velocity direction is mainly dependent on the distance to the object. As the optimal grasp pose is continuously updated, the human can always influence the position of the hand relative to the object, i.e., the human can withdraw the action, or move around the object to select different grasp configurations. Thus, the proposed semi-autonomy scheme supports the user in moving and orienting the robot towards the object, and automates the grasping process when it is triggered, as shown in Figure 5.12. The object detection is realized in combination with a scene analysis and uses the same method as integrated in the DLR Shared Autonomy Setup (Section 5.1, [33]).

5.6. Study 3 – Assisting Hand Positioning

This section presents Study 3, where the effect of assistance for hand positioning relative to the object (C1) is evaluated in the context of the DLR assistive robotics system. In this setup, the operator can only directly control three degrees of freedom of the robotic end effector and hence, is very restricted in the tasks he can perform with the robotic arm. Therefore, the semi-autonomous assistance is used to set the rotational degrees of freedom and enable the operator to choose his approach direction to the object according to his task. In comparison to the DLR shared autonomy system, this limitation in the human interface makes the the DLR assistive robotics system particularly suited for Study 3. We are especially interested in how the assistance influences the objective performance of the system and the workload of the human operator. As we focus on the reaching motion towards the object, it is very likely that the mapping between HMI and robot has an effect on these measures. Therefore, we do not vary the hand configurations as in the previous studies, but we vary the degree of distortion between commanded velocities of the user and executed velocities by the robot. We assume that if the subjective world model is incongruent with the perceived sensory data (i.e. the robot moves in a different direction than the commanded one), users need to rely more on assistance in order to grasp the object.

In summary, the main hypotheses for Study 3 are

H1 Performance increases when using assistance.

H2 Workload decreases when using assistance.

H3 A distorted mapping leads to a more significant effect of assistance.

The experimental setup is first described, followed by the summary of results and a discussion on the findings.

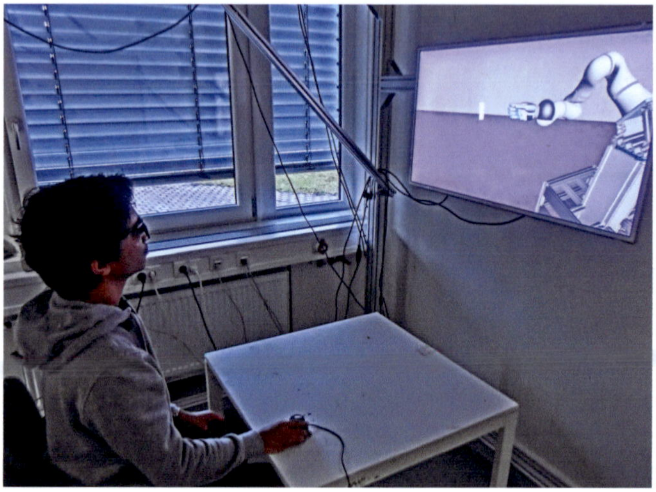

Figure 5.13.: Setup for Study 3 to evaluate the assistance for hand positioning with virtual fixtures.

5.6.1. Experimental Setup

In order to reduce the complexity of the system, the operator controls the velocity of the end effector of a simulated manipulator arm with a SpaceMouse [1]. Replacing an input based on bio-signals with a SpaceMouse allows for a faster setup (as no training for the bio-signals is needed) and a reduction of side effects that might occur from that training. Using a simulation instead of real-world experiments, minimizes the effect of errors from calibration and object detection in the real world and allows to explore the effect of assistance specifically. Thus, the chosen experimental setup leads to a high degree of experimental control, leading to a fair comparison of trials between the participants. The overall setup is shown in Figure 5.13, where a participant sits in front of the simulation visualized on a 3D TV[8]. In order to ensure depth perception during the task, participants wear active shutter glasses. As visualization software, the Instantplayer from Fraunhofer IGD[9] is used. The virtual scene resembles the physical robotic setup and consists of a DLR light-weight robot arm, the DLR-HIT Hand II, and a cylindric object that should be grasped. The scene is viewed on the screen from the perspective of a user sitting next to the robot base on the left, as shown in Figure 5.13.

[8] Samsung UE46ES6300 46inch 3-D LED monitor
[9] http://www.instantreality.org/

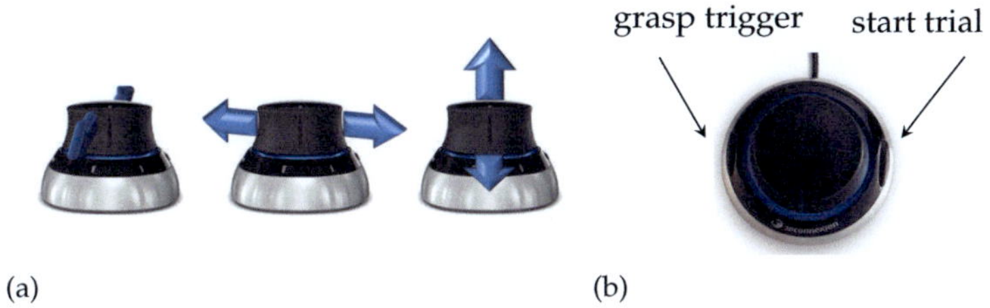

Figure 5.14.: Control of (a) the end effector velocity as well as (b) the grasp trigger with the SpaceMouse.

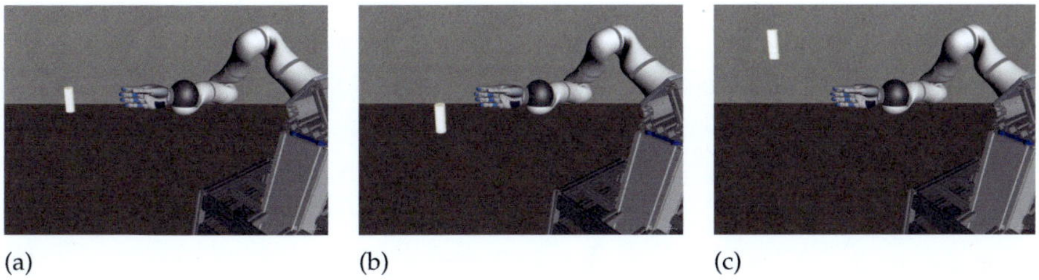

Figure 5.15.: Three different relative initial positions between the cylinder and the robotic hand. (a) "Easy": Motion in one axis required; (b) "Medium": Motion in two axis required; (c) "Difficult": Motion in three translational axis required.

Procedure

Participants were asked to grasp a virtual cylinder from the right side as fast as possible. They used the SpaceMouse to command the translational velocity of the DLR-HIT Hand II, and its buttons to start each trial and to trigger a grasp as shown in Figure 5.14. To vary the reaching motion towards the object, three different relative initial positions between the cylinder and the robotic hand were chosen as depicted in Figure 5.15. The difficulty of the positions increases from an "easy" position (Figure 5.15a), where the user needs to move the end effector only in one translational axis, to "difficult" (Figure 5.15c), where a motion in all three translational axis is required to grasp the object from the side. The object is not placed on a table, but instead is free floating in space. This does purposely not restrict the approach directions to the object although grasping from below was prevented.

The influence of mapping between user command and robot motion is additionally analyzed, as an HMI based on bio-signals often results in noisy data

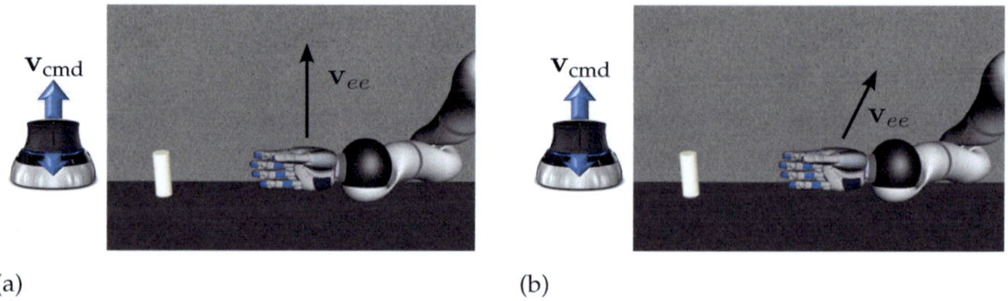

Figure 5.16.: Participants had to control the velocity of the end effector of the robot in two modes: (a) undistorted and (b) distorted.

and hence, in a distorted mapping between commanded velocity direction $\mathbf{v}_{\text{cmd}}$ and movement of the robot $\mathbf{v}_{ee}$. Therefore, all participants had to grasp the object in two modes, distorted and undistorted mapping:

$$\mathbf{v}_{ee} = \begin{cases} \mathbf{I} \cdot \mathbf{v}_{\text{cmd}} & \text{undistorted mapping} \\ \begin{pmatrix} \sigma_{xx} & \sigma_{xy} & \sigma_{xz} \\ \sigma_{yx} & \sigma_{yy} & \sigma_{yz} \\ \sigma_{zx} & \sigma_{zy} & \sigma_{zz} \end{pmatrix} \cdot \mathbf{v}_{\text{cmd}} & \text{distorted mapping.} \end{cases} \tag{5.1}$$

For the experimental setup in the user study, we choose $|\sigma_{i,j}| = 0.3$ for $i \neq j$ and $\sigma_{i,j} = 1$ for $i = j$ and the distoration matrix was normalized to prevent a scale of the magnitude of the user commandeed velocity. The signs of $\sigma_{i,j}$ are selected such that the distorted directions point away from the target object. Undistorted, the mapping between commanded velocity direction and movement of the robot match, as depicted exemplary for a velocity command in the z-axis in Figure 5.16a. In distorted mode, a command in one axis results in a movement of the robot in all three translational axis (Figure 5.16b).

Overall, three assistance conditions were tested: a no-assistance (manual) condition, and two conditions with assisted hand positioning. In the manual condition, the participant moves the end effector manually to the object by commanding translational velocities and the robot follows the commands according to the chosen mapping. When the end effector is within an area of 0.05 m from the surface of the object, the object color changes to green, and the user can trigger the grasp to successfully finish the grasp trial. In this mode, the fingers of the hand move to a pre-defined joint configuration reflecting the behavior of the hand in the physical setup. For both assistance conditions, the method of assistance is based on dynamic virtual fixtures as presented in Section 4.3.

Thus, the participant commands the translational velocities, the autonomous behavior selects a target grasp according to the intend of the user, and adapts the orientation of the end effector accordingly. The two conditions differ only in the distance between hand and object that is allowed upon triggering a grasp: in the condition "trigger-when-near", the end effector needs to enter the same area as in the "manual" condition, while in the condition "trigger-when-far", a distance of 0.13 m from the surface of the object is allowed. As soon as a grasp trigger is authorized, the object color changes to green as in the manual condition, and when the grasp is activated, the fingertips move to the planned contact points to form the grasp.

For each block of trials with 1 assistance condition and 1 mode of mapping, subjects had to grasp the object in all three poses in ascending difficulty with two repetitions of the complete cycle. Altogether, a number of 6 blocks of grasp trials (3 assistance conditions x 2 modes of mapping) and 36 total grasp trials (6 blocks x 6 trials) had to be completed. The order of the assistance conditions was systematically permuted to control for potential time effects like learning or fatigue. After completing 3 blocks of grasp trials with the undistorted mapping, the same order of assistance conditions was executed with the distorted mapping. Hence, all participants were already trained to use the system when they controlled it with the distorted mapping. The participant started each trial by pressing a button on the SpaceMouse, and he had then a maximum of 20 s to reach the object and trigger the grasp. After each block, parts of the NASA-TLX questionnaire [79] regarding mental and physical demand of the task, as well as a specific questionnaire to evaluate naturalness of interaction, interactivity, and consistency of mapping [186] were administered (given in Section A.3.1). The sense of presence was indicated by answering the item "How strong was your sense of presence in the virtual environment (no sense of presence: 0 % to sense of presence like in the real world: 100 %)". After three blocks of trials within one mode of mapping, participants evaluated the assistance conditions in terms of workload, consistency, and usefulness (Section A.3.2). Prior to the experiments, participants were informed about the objectives of the experiment and signed a consent form. Then, they were introduced in the setup, the experimental task, and the procedure. They explored the behavior of the system in a short training phase of 60 s in manual, undistorted mode.

Sample

All participants are technical staff and students from the Institute of Robotics and Mechatronics at DLR. 18 subjects participated in the user study with an

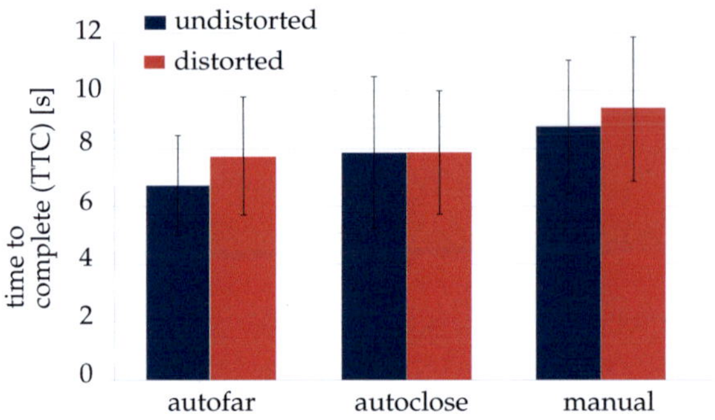

Figure 5.17.: Time to complete the task (TTC) in Study 3 for undistorted (blue) and distorted (red) control mode.

average age of $M = 26.7\,\text{y}$ ($SD = 2.6$). 11 participants were male and 7 female, all with a background of engineering, but none of them had previous experience with the setup. In fact, 12 of the participants do not work at all with CAD programs, and hence, had only limited to no experience with the SpaceMouse. The other 6 participants work on average 3.7 h per week with CAD programs. Nevertheless, analysis revealed that it had no influence on their performance during the study.

5.6.2. Results

All results are summarized in Table 5.6, and we start by giving the results of the objective data, namely time to complete the task (TTC), the overall translational movement of the end effector during the task, and the task success. Then, the subjective measures in terms of workload, sense of presence, consistency of interaction, and usefulness of the conditions are reported.

Time to Complete (TTC, Figure 5.17)
A repeated measures analysis of variance (ANOVA) with *assistance* (manual vs. trigger-when-near vs. trigger-when-far) as repeated measures was performed on the time to complete (TTC) each grasp, averaged over all object positions. For both control modes (undistorted and distorted), a highly significant main effect occurred ($Fs(2, 34) > 7.0; ps < 0.005$). Paired difference t-tests (two-sided) revealed a significantly shorter TTC for the distorted mode, when performing the task with assistance compared to no assistance ($M = 9.39\,\text{s}$). The effect was higher for assistance far ($M = 7.75\,\text{s}$, $t(17) > 2.88$, $p < 0.01$)

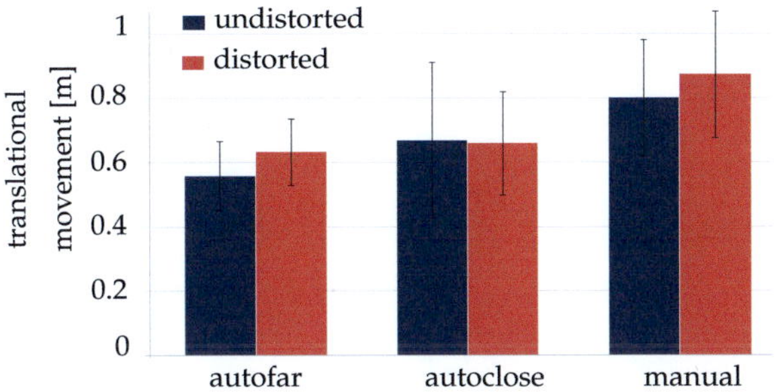

Figure 5.18.: Complete translational movement of the end effector in Study 3 for the undistorted (blue) and the distorted (red) control mode.

than for assistance near ($M = 7.87\,\text{s}$, $t(17) > 2.70$, $p < 0.015$). No effect between the assistance modes was found ($t(17) = 0.45$, n.s.). Interestingly, in the undistorted control mode, only the TTC of the trigger-when-far condition ($M = 6.73\,\text{s}$) is significantly shorter than the manual condition ($M = 8.77\,\text{s}$, $t(17) = 3.79$, $p < 0.005$). Comparing trigger-when-near to the manual condition revealed no significant effects ($M = 7.87\,\text{s}$, $t(17) = 1.84$, n.s.).

Translational Movement, Figure 5.18

The analysis revealed a significant main effect of the assistance for both the undistorted ($F(2, 34) = 9.96$, $p < 0.001$) and the distorted control mode ($F(2, 34) = 23.45$, $p < 0.001$). As expected, the effect is more significant for the distorted mode. In both control modes, the manual condition revealed significantly more translational movement ($M_{\text{undist}} = 0.80\,\text{m}$, $M_{\text{dist}} = 0.87\,\text{m}$) than the trigger-when-far condition ($M_{\text{undist}} = 0.56\,\text{m}$, $M_{\text{dist}} = 0.63\,\text{m}$, $ts(17) > 4.96$, $ps < 0.001$). In the distorted mode, also the trigger-when-near condition ($M = 0.66\,\text{m}$) reduced the translational movement significantly compared to the manual condition ($t(17) = 4.38$, $p < 0.001$). Between the assistance conditions, no significant effects could be found for both modes ($ts < 1.85$, n.s.).

Task Success

Repeated measures ANOVA revealed no significant effects regarding the task success ($Fs < 3.3$,n.s). Sometimes, participants could not finish the task within the 20 s in all conditions and modes. This occurred once or twice during the 12 grasps a participant had to execute. Only in distorted, manual control mode all grasp trials succeeded.

Table 5.6.: Overview of results of Study 3 with means (standard deviation)

	normal			distorted		
	Manual	Trigger-when-near	Trigger-when-far	Manual	Trigger-when-near	Trigger-when-far
TTC, Time to Complete [s]	8.8 (2.3)	7.9 (2.6)	6.7 (1.7)	9.4 (2.5)	7.9 (2.1)	7.7 (2.0)
Translational Movement [m]	0.8 (0.2)	0.7 (0.2)	0.6 (0.1)	0.9 (0.2)	0.7 (0.2)	0.6 (0.1)
Task Success [%]	99.1 (0.2)	97.2 (0.4)	97.2 (0.4)	100 (0)	94.4 (0.6)	94.4 (0.6)
Mental Workload (very low: 0; very high: 20)	5.7 (3.8)	4.5 (3.7)	3.3 (2.9)	7.2 (4.1)	5.0 (3.5)	4.6 (3.8)
Physical Workload (very low: 0; very high: 20)	3.6 (2.1)	3.1 (1.9)	2.3 (1.1)	4.5 (3.2)	3.2 (2.1)	2.8 (1.5)
Easy to Grasp (not true: 1; true: 7)	4.3 (1.6)	5.8 (1.0)	6.2 (0.7)	4.0 (1.8)	5.2 (1.2)	5.8 (0.9)
Sense of Presence [%]	60.0 (26.9)	58.9 (28.2)	60.0 (29.9)	57.8 (27.8)	60.8 (27.1)	60.3 (28.8)
Consistency (very low: 1; very high: 7)	5.8 (1.3)	6.3 (0.8)	6/0 (0.9)	5.0 (1.5)	5.5 (1.3)	6.0 (1.2)
Usefulness (very low: 1; very high: 7)	4.0 (1.5)	5.2 (1.2)	5.3 (1.5)	4.2 (1.5)	5.8 (0.8)	6.3 (0.7)

Workload

We monitored the mental and the physical workload of the participants separately. As shown by repeated measures ANOVA, both measures yield significant results, although the effect is stronger for the distorted mode. In undistorted control mode, the effect for mental workload is $F(2, 34) = 7.20, p < 0.005$, and for the physical workload $F(2, 34) = 7.27$, $p < 0.005$. The mental demand in the trigger-when-far condition ($M = 3.28$) is significantly lower than in the manual condition ($M = 5.67$, $t(17) = 3.72$, $p = 0.0017$) as well as in the trigger-when-near condition ($M = 4.50$, $t(17) = 2.93$, $p < 0.01$). There was no significant difference found between the assistance conditions ($t(17) = 1.5$, n.s.). The physical demand of the trigger-when-far condition ($M = 2.33$) is significantly lower than the manual condition ($M = 3.61, t(17) = 3.38, p < 0.005$), the other comparisons were not significant ($ts(17) < 2.6$, n.s.).

The effects for the distorted control mode show the same pattern. Mental demand reduced significantly comparing the manual condition ($M = 7.22$) to the trigger-when-far ($M = 4.56$, $t(17) = 3.76$, $p = 0.0016$) and the trigger-when-near condition ($M = 5.0$, $t(17) = 3.08$, $p = 0.007$). No significant difference between the assistance conditions could be found ($t(17) = 0.88$, n.s.). Also, the physical demand reduced from manual condition ($M = 4.50$) compared to the trigger-when-far condition ($M = 2.78$, $t(17) = 2.95$, $p < 0.01$), but not for the other conditions ($ts(17) < 2.85$, n.s.).

Additionally, participants rated the workload in the different conditions with the following question "In following condition, it was easy for me to grasp the object" (not true at all: 1; fully true: 7). In both control modes, highly significant effects are found ($F_{\text{undist}}(2, 34) = 16.86$, $F_{\text{dist}}(2, 34) = 14.64$, $ps < 0.001$). T-tests revealed that grasping with assistance is significantly easier for participants compared to the manual condition (Undistorted: $M_{\text{manual}} = 4.33$, $M_{\text{t-near}} = 5.83$, $M_{\text{t-far}} = 6.17$, $ts(17) > 3.79, p < 0.002$; Distorted: $M_{\text{manual}} = 4.0$, $M_{\text{t-near}} = 5.17$, $M_{\text{t-far}} = 5.83$, $ts(17) > 3.60, ps < 0.003$).

Consistency of Interaction

The consistency of interaction was rated with the question "In following condition, I understood when I could grasp the object" (not true at all: 1; fully true: 7). Interestingly, no main effect could be found in the undistorted control mode ($F(2, 34) = 3.04$, n.s.), but for the distorted mode ($F(2, 34) = 31.9$, $t(17) < 0.001$). Apparently, the assisted motion was more consistent to their expectation ($M_{\text{t-near}} = 5.5$, $M_{\text{t-far}} = 6.0$) than the distorted, manually operated motion ($M = 5.0, ts > 3.33, ps < 0.004$).

Usefulness

Finally, participants reported the usefulness ("Rate the usefulness of the conditions to grasp an object" (not at all: 1; very much: 7)). For both control modes, highly significant effects were revealed ($F_{\text{undist}}(2, 34) = 13.16$, $F_{\text{dist}}(2, 34) = 40.27$, $ps < 0.001$). In the undistorted control mode, the manual condition ($M = 4.0$) was rated less useful than the assistance conditions ($M_{\text{t-near}} = 5.17$, $M_{\text{t-far}} = 5.33$, $ts(17) > 3.17$, $ps < 0.006$). The same effect was reported in the distorted mode ($ts(17) > 4.42$, $ps < 0.001$). Additionally in this mode, the trigger-when-far condition was significantly more useful than the trigger-when-near condition ($t(17) = 4.12, p < 0.001$).

5.6.3. Discussion

In summary, all three hypotheses could be confirmed with the results of the user study. Performance significantly increased with assistance as shown with a decreased TTC and translational movement (H1 confirmed). The effect was significant for both assistance conditions for the undistorted mode, but only for the trigger-when-far condition in distorted mode. This leads to the conclusion that the assistance was especially necessary when participants were unfamiliar with the system. Once they learned to control the hand, even if the control was distorted, the trigger-when-far condition was more helpful to increase performance. No significant effect for the task success rate was found, but in all conditions, grasps failed from time to time. In manual control mainly because participants were not familiar with the system. For the assistance conditions, failures occured as the workspace of the arm was not taken into account when selecting the target grasp pose. Thus, the autonomy sometimes chose grasps poses for the hand, that the arm could not reach. The workload of participants decreased with the use of assistance, thus, we can support H2, especially for the trigger-when-far assistance. Also, H3 can be confirmed: the positive effects of assistance are stronger for the distorted control mode. TTC and translational movement reached a higher level of significance compared to the undistorted mode. Especially in the subjective measures of consistency and usefulness, participants found assistance more important in the distorted mode. There, the assisted motion is significantly more consistent to their expectation, in contrast to the undistorted mode. In terms of usefulness, only in distorted mode, a significant difference between the assistance conditions could be found. These findings stress the importance of assistance in the real setup, as signals obtained from a bio-HMI are often distorted.

5.7. Discussion

In all studies, the developed assistance supports the operator in finding and executing stable grasps in all the steps in human motor schemes for grasping: hand positioning, finger adjustment, and the actual grasp. He receives visual assistance and assistance with virtual fixtures for finding a hand pose relative to the object that leads to stable force closure grasps. Adequate contact positions on the object surface are computed online using reachable contact regions, and are displayed to the human operator to assist him in adjusting the finger position for grasping. The actual grasp is finally executed by the robotic hand when commanded by the human using a one degree of freedom hand input device that leads the fingers to the desired contact positions inside the contact regions. Nevertheless, the overall level of autonomy evaluated in this section is moderate, to avoid "out of the loop" phenomena: the human is in control of the arm motion, thus enabling his active participation in the prehension process.

The assistance is first evaluated with two user studies (Study 1 and 2) that also evaluate the influence of different kinematics of the robotic end effector, and therefore the influence of mapping in the performance of telepresence systems with assistance. In both studies, the kinematics of the end effector was modified to compare a natural looking (NL) hand with a non-natural looking (NNL) hand by changing the position of the thumb. Both studies revealed a better grasping performance with assistance for both hand configurations, and also reported a decrease in workload for the operator. Then, the influence of mapping is considered specifically in Study 3, where an undistorted velocity mapping is compared to a distorted mode. In general, the study confirmed a higher workload of the user with a distorted mapping which could be significantly reduced using assistance for hand positioning. Also, the grasping performance increased with assistance for both mapping modes.

Taking a closer look into the assistance effects caused by the appearance and functionality of the end effector, Study 1 and 2 reported two different results. In Study 1, the assistance effect was stronger for the NNL hand, which could not be confirmed in Study 2. The unnatural position of the thumb of the NNL hand in Study 1 lead to an increased workspace that was better suited for the grasping tasks. Although the mapping between the hands was different, the reaching motion was almost the same for all prehension tasks as all objects were placed on the same position. Thus, the operators could focus on the presented assistance for the finger adjustment, and the better thumb functionality allowed a better performance for the NNL hand. In contrast, one focus of Study 2 was the assistance for reaching motions, which were changed by using different target poses for the object. The mismatch in mapping between the human and the NNL hand triggered a learning process that resulted in increased attention towards the virtual scene and higher workload for the operator. The assistance for hand pose did not strongly influence the performance of the task, as the mapping for the hand needed to be learned first. In Study 3, the positive assistance effect was stronger with a distorted mapping mode which corresponds to the findings in Study 1. It leads also to the conclusion, that a pure visual assistance might not be sufficient to assist hand positioning if the mapping is unintuitiv. This can result from the higher workload which was reported in Study 2 and 3 for a change in mapping. Then, users have to process additional information, in contrast to an assistance with virtual fixtures for hand positioning.

In Study 1 and 2, the perceived sense of presence depends on the mapping between the human hand and the robotic end effector, but in Study 3 this effect could not be confirmed. In Study 1, the NL hand was tougher to operate as it looks like an anthropomorphic hand but has different capabilities. This mismatch resulted in a slightly lower sense of presence for the NL hand than the NNL hand. In Study 2, the thumb position of the NNL hand was so awkward for the operators that the sense of presence was higher for the NL hand in general. Nevertheless, visual assistance increased the sense of presence for the NNL hand.

Summarizing, the performance of operators in telepresence systems with robotic hands as end effectors can be increased with the presented assistance, while also decreasing the workload. Results reveal a high influence of the shape of the end effector, and suggest that an increase in functionality (as obtained for the NNL hand in Study 1) can counterbalance a decrease in "naturalness". In fact, this decrease can even be helpful to visualize differences between human and robotic hands, which triggers a learning process for the human that can be later beneficial for obtaining a higer involvement of the user with the system. Study 2 showed that the increased task difficulty (the object had to be liftable) lead to a decrease in task success rates. Obviously, the dynamic behavior of the hand was not visualized in a transparent way and was not easily interpretable by the participants. In contrast, Study 3 revealed a highly positive effect of task usefulness, robot behavior expectations, and grasping for distorted mapping although they had already familiarized with the system.

6 Conclusion and Future Works

6.1. Conclusion

This thesis presents a new concept for assisting the operator of a complex telepresence system in grasping and manipulation tasks. Assistance is crucial in these tasks as they are error-prone and it is exhausting for the human operator in current telepresence systems. For designing such assistance, we consider two major aspects:

Handling: The human operator needs to be able to use and apply his task knowledge, as well as his ability to solve problems in unexpected situations. Furthermore, the system should be usable without expert knowledge about the robotic system.

Interface: Additional information provided by the autonomous functions should be displayed appropriately to the human operator.

In comparison to state of the art approaches for assistance, we take these challenges into account by keeping the high level of immersion of the human in telepresence systems. This enables a low workload for the operator, keeps him concentrated on the task, and hence, he is able to apply his task knowledge. It also avoids "out-of-the-loop" phenomena, as described by [58], which can lead to incorrect reactions of the human, particularly in the case of system or task failure. We allow the human to actively participate in the prehension process as he continuously controls the hand pose relative to the object. Additionally, the operator experiences multi-modal feedback, including haptic force and stereo visual feedback. The visual feedback of the remote environment is augmented such that the operator is informed about the current system capabilities and the current state of the autonomous functions.

After analyzing human prehension, we also identified three challenges for grasping and manipulation in teleoperation systems:

C1 - hand positioning
Setting up a position and orientation of the hand relative to the object.

C2 - finger adjustment
Finding contact points and the corresponding finger joint configurations for a stable grasp.

C3 - actual grasp
Moving into contact with all fingers, applying suitable contact forces, and ensuring grasp stability.

These three challenges are met by developing new assistance methods for hand positioning (C1) and grasp planning (C2), which also improve the execution of the actual grasp with the robotic hand (C3).

The developed grasp planning is based on the concept of independent contact regions, which makes it possible to compute contact areas instead of single contact points. Hence, a certain robustness to positioning errors in the real robotic system is inherently built in. Additionally, the method makes it possible to plan contact points and contact forces simultaneously, and in contrast to state of the art approaches, it takes into account the realistic contact forces that the robotic hand can apply on the object surface. This allows to plan grasps that consider the workspace boundaries and the torque limits of the hand. The grasps are computed only based on the hand pose relative to object, which is given by the human operator. A low computation time ensures that the grasp planner can cope with the movement speed of the human.

The assistance for hand positioning combines a dense database of grasps with the concept of virtual fixtures. This ensures that the human can choose the approach direction to the object freely and is not restricted to a few pre-programmed grasps. The dense database of grasps is based on the developed grasp planning and stores all precision grasps for a hand-object combination. Within this database and based on the currently commanded velocity directions, a grasp is chosen. It represents the goal for the virtual fixture, which guides the hand movement towards it. The target grasp is adaptively recomputed in every time step of the reaching motion, which means that the human can always choose a different grasp, change the approach direction, or even retract from the object.

Finally, the presented concept of assistance for grasping and manipulation in telepresence systems is evaluated with psycho-physical user studies. They take into account objective and subjective measures, like time to complete the task and the perceived level of presence, to examine system performance. The studies show that each developed assistance reduces the workload of the operator statistically significantly while simultaneously improving the performance of grasping tasks. The perceived level of immersion of the operator into the remote environment is not at the same level as in the telepresence system without assistance, but it is even increased by the provided assistance.

6.2. Future Works

Shared autonomy in telerobotics is a tool to improve system performance and its extensive use in scenarios such as the ones provided in the DARPA robotics challenge, shows its importance. The method of integrating assistance in a complex telepresence system while keeping the high immersion of the operator into the remote environment has a high potential to realize an all-purpose telerobotic system. The presented approach can serve as the basis for such a system. Thereby, it is important that the flexibility to react to unforeseen events provided by direct teleoperation is still given with telerobotic systems with a higher level of autonomy. In this thesis, a grasping assistance is presented which is limited to precision grasps. As a next step, the grasping and manipulation assistance should therefore be enhanced to consider other grasp types, like power grasps, or in-hand manipulation. Similar to the graspability map, a dense database of power grasps or of in-hand manipulation tasks can be included within the framework. The current approach for precision grasps relies

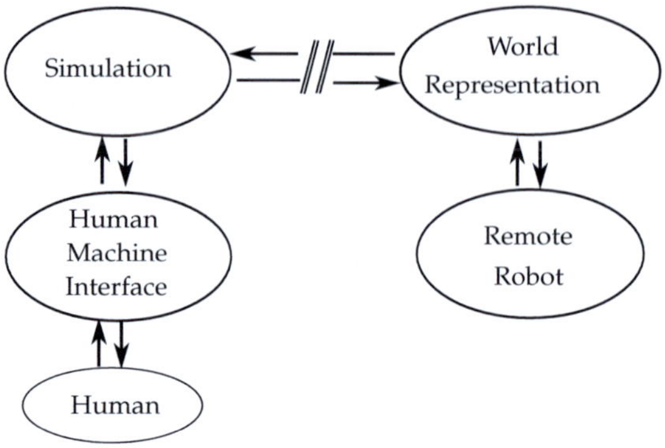

Figure 6.1.: Towards a higher level of autonomy in telerobotic systems.

on the analytic force closure criterion in combination with the computation of contact forces. For power grasps, this concept is computationally more expensive which can lead to high calculation times. Therefore, the use of a physics simulation to calculate the database should be considered, although the grasps selected by current physics simulations do not necessarily lead to good grasps in the real world. In [242], we proposed a method to plan in-hand manipulation tasks, that can be incorporated in the proposed shared autonomy framework. In comparison to commanding power or precision grasps, the selection of in--hand manipulation tasks demands further investigation in the interface to the human operator, which was not considered in the publication. The interface to the human also needs to be extended such that he is able to choose between the provided grasp types.

Another restriction, which should be overcome by future works, is the assumption of known objects. Finding grasps for unknown objects is tough especially if visual information is limited to only one side of the object. Then, some assumption has to be made to estimate the missing parts of the object in order to be able to place the fingers of a robotic hand on the object surface such that the grasp is stable. We would combine the approach of fitting basic shapes, like cylinders and spheres [134], and exploring the unknown object with the robotic hand. An iterative refinement of the basic shape by letting the human move the robotic hand around the object or even applying forces on the surface can lead to an accurate estimate of the object.

Finally, a system that enables different assistance modes and a method to handle unknown objects opens the way to realize the concept of teleprogramming in a new, flexible manner. As shown in Figure 6.1, in this concept, the human is coupled to a simulation environment which communicates with the world representation of the remote robot. Instead of selecting pre-programmed skills or behaviors, as proposed in state of the art teleprogramming, the coupling to a physical simulation, which can be updated by the sensor measurements of the remote robot and can incorporate unknown objects, restores the flexibility to cope with unforeseen situations. Of course, the concept relies on an accurate simulation of the physical properties of the environment and a correct estimate of the world state of the robot.

A
User Study Questionnaires

A.1. Study 1

A.1.1. Questionnaire "TLX"

Zwischenfragebogen - Arbeitsbelastung (NASA-TLX)

Vp-Nr.: ____________ Bedingung: |AH| |NH| - |K| |D| |GR| (vom Versuchsleiter auszufüllen)

Die folgenden Fragen beziehen sich auf die Arbeitsbelastung im letzten Durchgang.

Geistige Anforderungen Wie viel geistige Anstrengung war bei der Informationsaufnahme und bei der Informationsverarbeitung erforderlich? War die Aufgabe einfach oder komplex?

sehr wenig/ einfach — sehr viel/ /komplex

Körperliche Anforderungen Wie viel körperliche Aktivität war erforderlich? War die Aufgabe einfach oder anstrengend?

sehr wenig/ einfach — sehr viel/ anstrengend

Zeitliche Anforderungen Wie viel Zeitdruck empfanden Sie hinsichtlich der Häufigkeit oder dem Takt mit dem Aufgaben oder Aufgabenelemente auftraten? War die Abfolge langsam oder geruhsam oder schnell und hektisch?

Gering/ sehr langsam — hoch/ sehr schnell

Ausführung der Aufgaben Wie erfolgreich haben Sie Ihrer Meinung nach die vom Versuchsleiter (oder Ihnen selbst) gesetzten Ziele erreicht? *(1= Perfekt; 20 = Fehlschlag)*

Perfekt — **Fehlschlag**

Anstrengung Wie sehr mussten Sie sich anstrengen, um Ihren Grad an Aufgabenerfüllung zu erreichen?

sehr wenig — sehr viel

Frustration Wie unsicher, entmutigt, irritiert, gestresst und verärgert (versus sicher, bestätigt, zufrieden, entspannt und zufrieden mit sich selbst) fühlten Sie sich während der Aufgabe?

sehr wenig — sehr viel

Beanspruchungsstruktur im Experiment (NASA-TLX)

Geben Sie bitte an, welche relativen Bedeutungen für die empfundene Gesamtbeanspruchung in allen Durchgängen die sechs Beanspruchungsdimensionen *Geistige Anforderungen, Körperliche Anforderungen, Zeitliche Anforderungen, Ausführung der Aufgaben, Anstrengung und Frustration* für Sie hatten.
Im Folgenden werden jeweils zwei der sechs Beanspruchungsdimensionen in verschiedenen Kombinationen gegenübergestellt. Geben Sie jeweils an, welche Beanspruchungsdimension für die Gesamtbeanspruchung, die Sie empfunden haben, bedeutsamer war. Es geht also nicht darum, wie hoch die Beanspruchung in den einzelnen Dimensionen war, sondern wie wichtig die jeweilige Dimension für das Gesamtempfinden war!
Bsp.: Wenn die geistigen Anforderungen, die die Aufgabe gestellt hat für die empfundene Gesamtbeanspruchung bedeutsamer waren als die Anstrengung, die Sie aufbringen mussten, dann kreuzen Sie wie folgt an:

Anstrengung		X	Geistige Anforderungen

Anstrengung			Ausführung der Aufgabe
Anstrengung			Frustration
Ausführung der Aufgabe			Frustration
Ausführung der Aufgabe			Geistige Anforderungen
Ausführung der Aufgabe			Zeitliche Anforderungen
Frustration			Geistige Anforderungen
Frustration			Körperliche Anforderungen
Frustration			Zeitliche Anforderungen
Geistige Anforderungen			Anstrengung
Geistige Anforderungen			Körperliche Anforderungen
Körperliche Anforderungen			Anstrengung
Körperliche Anforderungen			Ausführung der Aufgabe
Zeitliche Anforderungen			Anstrengung
Zeitliche Anforderungen			Geistige Anforderungen
Zeitliche Anforderungen			Körperliche Anforderungen

A.1.2. Questionnaire "Hands"

Zwischenfragebogen – Hände

Vp-Nr.:____________ Bedingung: |AH| |NH| – (vom Versuchsleiter auszufüllen)

Die folgenden Fragen beziehen sich auf die Roboterhand, die in den vorangegangenen drei Aufgabenblöcken verwendet wurde. Bitte beantworten Sie alle Fragen, es gibt keine „richtigen" oder „falschen" Antworten, was zählt, ist Ihre Meinung.

1. Das Verhalten der Roboterhand entsprach stets meiner Erwartung.

trifft gar nicht zu — trifft völlig zu

2. Wie natürlich erschien Ihnen die Interaktion mit der Roboterhand in der entfernten Umgebung?

sehr unnatürlich — sehr natürlich

3. Ich fand es sehr mühsam, mit der Hand zu arbeiten.

trifft gar nicht zu — trifft völlig zu

4. Ich hatte das Gefühl, die Roboterhand sei meine eigene.

trifft gar nicht zu — trifft völlig zu

A.1.3. Questionnaire "Attrak-Diff"

Zwischenfragebogen – Gebrauchstauglichkeit (AttrakDiff: PQ)

Vp-Nr.:____________ Bedingung: |AH| |NH| - |D| |GR| (vom Versuchsleiter auszufüllen)

Nachfolgend finden Sie Wortpaare, mit deren Hilfe Sie die Assistenzvariante bewerten können. Sie stellen jeweils extreme Gegensätze dar, zwischen denen eine Abstufung möglich ist.

Beziehen sie sich in diesem Fragebogen bitte nur auf die vorangegangene Versuchsbedingung.

Denken Sie nicht lange über die Wortpaare nach, sondern geben Sie bitte die Einschätzung ab, die Ihnen spontan in den Sinn kommt. Kreuzen Sie bitte immer eine Antwort an. Denken Sie daran, dass es keine "richtigen" oder "falschen" Antworten gibt - nur Ihre persönliche Meinung zählt!

Die Assistenz in dieser Bedingung war:

Kompliziert								Einfach
Unpraktisch								Praktisch
Voraussagbar								Unberechenbar
Verwirrend								Übersichtlich
Direkt								Umständlich
Handhabbar								Widerspenstig

A.1.4. Final Questionnaire

Abschlussfragebogen

Vp-Nr.:_________________ Bedingung______________ (vom Versuchsleiter auszufüllen)

Assistenzvarianten:

1. In folgender Bedingung war es mir möglich, schnell einen stabilen Griff zu finden:

a)

Keine Assistenz	☐	☐	☐	☐	☐	☐	☐
	trifft gar nicht zu						trifft völlig zu

b)

Dichotome Rückmeldung	☐	☐	☐	☐	☐	☐	☐
	trifft gar nicht zu						trifft völlig zu

c)

Anzeigen von Greifregionen	☐	☐	☐	☐	☐	☐	☐
	trifft gar nicht zu						trifft völlig zu

2. Um einen möglichst präzisen Griff auszuführen, war folgende Bedingung am hilfreichsten:

a)

Keine Assistenz	☐	☐	☐	☐	☐	☐	☐
	trifft gar nicht zu						trifft völlig zu

b)

Dichotome Rückmeldung	☐	☐	☐	☐	☐	☐	☐
	trifft gar nicht zu						trifft völlig zu

c)

Anzeigen von Greifregionen	☐	☐	☐	☐	☐	☐	☐
	trifft gar nicht zu						trifft völlig zu

3. Auf welche Szene haben Sie in den folgenden Bedingungen stärker geachtet?

a)

Ohne Assistenz	☐	☐	☐	☐	☐	☐	☐
	nur auf die virtuelle Szene			auf beide Szenen gleich stark			nur auf die reale Szene

b)

Bei dichotomer Rückmeldung	☐	☐	☐	☐	☐	☐	☐
	nur auf die virtuelle Szene			auf beide Szenen gleich stark			nur auf die reale Szene

c)

Bei angezeigten Greifregionen	☐	☐	☐	☐	☐	☐	☐
	nur auf die virtuelle Szene			auf beide Szenen gleich stark			nur auf die reale Szene

4. In folgender Bedingung war ich mir schon mit dem Zugreifen sicher, dass mein Griff stabil ist.

a)

Ohne Assistenz							
	trifft gar nicht zu						trifft völlig zu

b)

Bei dichotomer Rückmeldung							
	trifft gar nicht zu						trifft völlig zu

c)

Bei angezeigten Greifregionen							
	trifft gar nicht zu						trifft völlig zu

5. Folgende Assistenzvariante war unnötig komplex.

a)

Dichotome Rückmeldung							
	trifft gar nicht zu						trifft völlig zu

b)

Anzeigen von Greifregionen							
	trifft gar nicht zu						trifft völlig zu

6. Ich kann mir vorstellen, dass die meisten Leute schnell lernen würden, mit der Assistenzvariante zu arbeiten.

a)

Mit dichotomer Rückmeldung							
	trifft gar nicht zu						trifft völlig zu

b)

Mit angezeigten Greifregionen							
	trifft gar nicht zu						trifft völlig zu

7. Ich halte die Assistenzvariante für zu unvorhersehbar.

a)

Dichotome Rückmeldung							
	trifft gar nicht zu						trifft völlig zu

b)

Anzeigen von Greifregionen							
	trifft gar nicht zu						trifft völlig zu

8. Ich fand es sehr mühsam, mit der Assistenzvariante zu arbeiten.

a)

Mit dichotomer Rückmeldung							
	trifft gar nicht zu						trifft völlig zu

b)

Mit angezeigten Greifregionen							
	trifft gar nicht zu						trifft völlig zu

9. Durch die Assistenz fühlte ich mich sehr sicher bei der Ausführung der Aufgabe.

a)

Mit dichotomer Rückmeldung							

trifft gar nicht zu … trifft völlig zu

b)

Mit angezeigten Greifregionen							

trifft gar nicht zu … trifft völlig zu

10. Bei der Assistenzvariante mit der **dichotomen Rückmeldung** gab es Situationen, in denen ich nicht wusste, meinen Griff verändern musste, um ihn zu verbessern.

trifft gar nicht zu … trifft völlig zu

11. Die Assistenzvariante mit der **dichotomen Rückmeldung** war vollkommen ausreichend, um einen stabilen Griff zu finden.

trifft gar nicht zu … trifft völlig zu

12. Bei der Assistenzvariante mit den **angezeigten Greifregionen** fand ich es schwierig, die Position der Greifregionen nachzuvollziehen.

trifft gar nicht zu … trifft völlig zu

13. Bei der Assistenzvariante mit den **angezeigten Greifregionen** gab es Situationen, in denen ich keinen Überblick mehr hatte und nicht mehr wusste, was die optimale Griffposition ist.

trifft gar nicht zu … trifft völlig zu

14. Sonstige Bemerkungen zu den Assistenzvarianten:

____

____

____

Hände: (Fotos)

15. Um möglichst schnell einen stabilen Griff zu finden, halte ich folgende Hand für geeignet:

a)

Mit alter Position des Daumens							
	trifft gar nicht zu						trifft völlig zu

b)

Mit neuer Position des Daumens							
	trifft gar nicht zu						trifft völlig zu

16. Wie natürlich erschien Ihnen die Interaktion mit der Roboterhand in der entfernten Umgebung?

a)

Bei alter Position des Daumens							
	sehr unnatürlich						sehr natürlich

b)

Bei neuer Position des Daumens							
	sehr unnatürlich						sehr natürlich

17. Die Position des Daumens hat mich bei der Interaktion stark gestört.

a)

Die alte Position des Daumens							
	trifft gar nicht zu						trifft völlig zu

b)

Die neue Position des Daumens							
	trifft gar nicht zu						trifft völlig zu

18. Nach einer kurzen Eingewöhnung habe ich nicht mehr über die Position des Daumens nachgedacht.

a)

Bei alter Position des Daumens							
	trifft gar nicht zu						trifft völlig zu

b)

Bei neuer Position des Daumens							
	trifft gar nicht zu						trifft völlig zu

19. Ich hatte den Eindruck, dass es eine zeitliche Verzögerung zwischen der Bewegung meiner Hand und der Bewegung der Roboterhand gab.

a)

Bei alter Position des Daumens							
	trifft gar nicht zu						trifft völlig zu

b)

Bei neuer Position des Daumens							
	trifft gar nicht zu						trifft völlig zu

20. Ich hatte den Eindruck, dass meine Handbewegung und die Bewegung der Roboterhand räumlich nicht übereinstimmten.

a)

Bei alter Position des Daumens							
	trifft gar nicht zu						trifft völlig zu

b)

Bei neuer Position des Daumens							
	trifft gar nicht zu						trifft völlig zu

21. Sonstige Bemerkungen zu den Händen:

____

____

____

22. Ich arbeite häufig mit CAD-Anwendungen

trifft gar nicht zu						trifft völlig zu

Alter: ____________ Jahre

Geschlecht: m w (bitte ankreuzen)

Beruflicher Hintergrund: ______________________________

A.2. Study 2

A.2.1. Questionnaire "TLX"

Zwischenfragebogen - Fragebogen zur Arbeitsbelastung (NASA-TLX)

Vp-Nr.: ____________________ Bedingung: _______ (vom Versuchsleiter auszufüllen)

Die folgenden Fragen beziehen sich auf die Arbeitsbelastung im letzten Durchgang.

Geistige Anforderungen Wie viel geistige Anstrengung war bei der Informationsaufnahme und bei der Informationsverarbeitung erforderlich? War die Aufgabe einfach oder komplex?

sehr wenig/ einfach	sehr viel/ /komplex

Körperliche Anforderungen Wie viel körperliche Aktivität war erforderlich? War die Aufgabe einfach oder anstrengend?

sehr wenig/ einfach	sehr viel/ anstrengend

Zeitliche Anforderungen Wie viel Zeitdruck empfanden Sie hinsichtlich der Häufigkeit oder dem Takt mit dem Aufgaben oder Aufgabenelemente auftraten? War die Abfolge langsam oder geruhsam oder schnell und hektisch?

Gering/ sehr langsam	hoch/ sehr schnell

Ausführung der Aufgaben Wie erfolgreich haben Sie Ihrer Meinung nach die vom Versuchsleiter (oder Ihnen selbst) gesetzten Ziele erreicht? *(1= Perfekt; 20 = Fehlschlag)*

Perfekt	**Fehlschlag**

Anstrengung Wie sehr mussten Sie sich anstrengen, um Ihren Grad an Aufgabenerfüllung zu erreichen?

sehr wenig	sehr viel

Frustration Wie unsicher, entmutigt, irritiert, gestresst und verärgert (versus sicher, bestätigt, zufrieden, entspannt und zufrieden mit sich selbst) fühlten Sie sich während der Aufgabe?

sehr wenig	sehr viel

Beanspruchungsstruktur im Experiment (NASA-TLX)

Geben Sie bitte an, welche relativen Bedeutungen für die empfundene Gesamtbeanspruchung in allen Durchgängen die sechs Beanspruchungsdimensionen *Geistige Anforderungen, Körperliche Anforderungen, Zeitliche Anforderungen, Ausführung der Aufgaben, Anstrengung und Frustration* für Sie hatten.

Im Folgenden werden jeweils zwei der sechs Beanspruchungsdimensionen in verschiedenen Kombinationen gegenübergestellt. Geben Sie jeweils an, welche Beanspruchungsdimension für die Gesamtbeanspruchung, die Sie empfunden haben, bedeutsamer war. Es geht also nicht darum, wie hoch die Beanspruchung in den einzelnen Dimensionen war, sondern wie wichtig die jeweilige Dimension für das Gesamtempfinden war!

Bsp.: Wenn die geistigen Anforderungen, die die Aufgabe gestellt hat für die empfundene Gesamtbeanspruchung bedeutsamer waren als die Anstrengung, die Sie aufbringen mussten, dann kreuzen Sie wie folgt an:

Anstrengung		X	Geistige Anforderungen

Anstrengung			Ausführung der Aufgabe
Anstrengung			Frustration
Ausführung der Aufgabe			Frustration
Ausführung der Aufgabe			Geistige Anforderungen
Ausführung der Aufgabe			Zeitliche Anforderungen
Frustration			Geistige Anforderungen
Frustration			Körperliche Anforderungen
Frustration			Zeitliche Anforderungen
Geistige Anforderungen			Anstrengung
Geistige Anforderungen			Körperliche Anforderungen
Körperliche Anforderungen			Anstrengung
Körperliche Anforderungen			Ausführung der Aufgabe
Zeitliche Anforderungen			Anstrengung
Zeitliche Anforderungen			Geistige Anforderungen
Zeitliche Anforderungen			Körperliche Anforderungen

A.2.2. Questionnaire “Attrak-Diff”

Zwischenfragebogen – Gebrauchstauglichkeit (AttrakDiff: PQ)

Vp-Nr.:____________ Bedingung: ______________(vom Versuchsleiter auszufüllen)

Nachfolgend finden Sie Wortpaare, mit deren Hilfe Sie die Assistenzvariante bewerten können. Sie stellen jeweils extreme Gegensätze dar, zwischen denen eine Abstufung möglich ist.

Beziehen sie sich in diesem Fragebogen bitte nur auf die vorangegangene Versuchsbedingung.

Denken Sie nicht lange über die Wortpaare nach, sondern geben Sie bitte die Einschätzung ab, die Ihnen spontan in den Sinn kommt. Kreuzen Sie bitte immer eine Antwort an. Denken Sie daran, dass es keine "richtigen" oder "falschen" Antworten gibt - nur Ihre persönliche Meinung zählt!

Die Assistenz in dieser Bedingung war:

Kompliziert								Einfach
Unpraktisch								Praktisch
Voraussagbar								Unberechenbar
Verwirrend								Übersichtlich
Direkt								Umständlich
Handhabbar								Widerspenstig

Wie stark war Ihr Präsenzempfinden in der virtuellen Umgebung (0% = keine Präsenz bis 100% = das Präsenzempfinden in der realen Welt)

Ich war im letzten Durchgang motiviert, eine gute Leistung zu erzielen.

trifft gar nicht zu						trifft völlig zu

A.2.3. Questionnaire "Immersion"

Immersive Tendenz

Mit den folgenden Fragen soll ermittelt werden, wie stark Sie sich im Allgemeinen in Ereignisse hineinversetzen (immersive tendency).

1. Lassen Sie sich leicht tief in Spiel- oder Fernsehfilme hineinziehen?

1	2	3	4	5	6	7
Nein			weder noch			Ja

2. Sind Sie manchmal so sehr in eine Fernsehsendung oder in ein Buch vertieft, dass andere Menschen nur schwer Ihre Aufmerksamkeit auf sich ziehen können?

1	2	3	4	5	6	7
nie			weder noch			sehr oft

3. Waren Sie jemals so sehr in einen Film vertieft, dass Ihnen die Dinge, die um Sie herum passierten, nicht mehr bewusst waren?

1	2	3	4	5	6	7
Nein, nie			weder noch			Ja, sehr oft

4. Wie oft identifizieren Sie sich stark mit den Charakteren einer Geschichte?

1	2	3	4	5	6	7
nie			weder noch			sehr oft

5. Fühlten Sie sich jemals so sehr in ein Computerspiel verwickelt, dass Sie eher das Gefühl hatten, Sie seien Teil des Spiels, als dass Sie nur einen Joystick bewegen und einen Bildschirm beobachten?

1	2	3	4	5	6	7
Nein, nie			weder noch			Ja, sehr oft

6. Wenn Sie sich Sportübertragungen ansehen: Sind Sie bisweilen so sehr in ein Spiel vertieft, dass Sie wie einer der Spieler handeln?

1	2	3	4	5	6	7
Nein, nie			weder noch			Ja, sehr oft

7. Werden Sie manchmal so sehr in einen Tagtraum hineingezogen, dass Sie sich der Dinge um Sie herum nicht mehr bewusst sind?

1	2	3	4	5	6	7
Nein, nie			weder noch			Ja, sehr oft

8. Haben Sie manchmal Träume, die so real sind, dass Sie sich beim Erwachen desorientiert fühlen?

1	2	3	4	5	6	7
Nein, nie			weder noch			Ja, sehr oft

9. Hat jemals eine Jagd- oder Kampfszene in Film oder Fernsehen Aufregung bei Ihnen ausgelöst?

1	2	3	4	5	6	7
Nein, nie			weder noch			Ja, sehr oft

10. Hat Sie jemals etwas in einer Fernsehsendung oder in einem Spielfilm geängstigt?

1	2	3	4	5	6	7
Nein, nie			weder noch			Ja, sehr oft

11. Ist es Ihnen jemals passiert, dass Sie nach einem beängstigenden Film lange Zeit besorgt oder verängstigt waren?

1	2	3	4	5	6	7
Nein, nie			weder noch			Ja, sehr oft

12. Sind Sie manchmal so mit einer Sache beschäftigt, dass Sie die Zeit vergessen?

1	2	3	4	5	6	7
Nein, nie			weder noch			Ja, sehr oft

A.2.4. Questionnaire "SimulatorSickness"

Simulator Sickness Questionnaire (nach Mehlitz, 2004)

Bitte füllen Sie folgenden Fragebogen aus, der verschiedene Symptome aufführt, die in Simulationsumgebungen auftreten können. Die Daten werden anonymisiert, d.h. ohne eine mögliche Zuordnung zu Ihrer Person, gespeichert. Falls Sie Fragen zu den einzelnen Symptomen haben, sprechen Sie bitte Ihren Testleiter an.

Symptom	**Ausprägung**			
(1) Allgemeines Unwohlsein	( ) nicht vorhanden	( ) etwas	( ) deutlich	( ) sehr stark
(2) Müdigkeit	( ) nicht vorhanden	( ) etwas	( ) deutlich	( ) sehr stark
(3) Kopfschmerzen	( ) nicht vorhanden	( ) etwas	( ) deutlich	( ) sehr stark
(4) Überanstrengte Augen	( ) nicht vorhanden	( ) etwas	( ) deutlich	( ) sehr stark
(5) Schwierigkeiten beim Scharfsehen	( ) nicht vorhanden	( ) etwas	( ) deutlich	( ) sehr stark
(6) Erhöhter Speichelfluss	( ) nicht vorhanden	( ) etwas	( ) deutlich	( ) sehr stark
(7) Schwitzen	( ) nicht vorhanden	( ) etwas	( ) deutlich	( ) sehr stark
(8) Übelkeit	( ) nicht vorhanden	( ) etwas	( ) deutlich	( ) sehr stark
(9) Konzentrationsschwierigkeiten	( ) nicht vorhanden	( ) etwas	( ) deutlich	( ) sehr stark
(10) Druckgefühl im Kopf	( ) nicht vorhanden	( ) etwas	( ) deutlich	( ) sehr stark
(11) Getrübtes Sehen	( ) nicht vorhanden	( ) etwas	( ) deutlich	( ) sehr stark
(12) Benommenheit bei geöffneten Augen	( ) nicht vorhanden	( ) etwas	( ) deutlich	( ) sehr stark
(13) Benommenheit bei geschlossenen Augen	( ) nicht vorhanden	( ) etwas	( ) deutlich	( ) sehr stark
(14) Schwindelgefühl	( ) nicht vorhanden	( ) etwas	( ) deutlich	( ) sehr stark
(15) Wahrnehmung des Magens	( ) nicht vorhanden	( ) etwas	( ) deutlich	( ) sehr stark
(16) Aufstoßen	( ) nicht vorhanden	( ) etwas	( ) deutlich	( ) sehr stark

A.2.5. Final Questionnaire

VP-Nr.:

Abschlussfragebogen

Alter: ____________ Jahre

Geschlecht: m w (bitte ankreuzen)

Beruflicher Hintergrund: ______________________________

Ich kenne das System.
ja ☐ nein ☐

Ich habe bereits vor dieser Studie mit dem System gearbeitet.
ja ☐ nein ☐

Wenn ja, wie oft?______________________________________

1. In folgender Bedingung war es mir möglich, schnell einen ausreichend guten Griff zu finden.

a.

	trifft gar nicht zu						trifft völlig zu
Kombination Assistenz zur Handposition, Assistenz zur Griffstabilität (permanent)							

b.

	trifft gar nicht zu						trifft völlig zu
Kombination Assistenz zur Handposition, Assistenz zur Griffstabilität (adaptiv)							

c.

	trifft gar nicht zu						trifft völlig zu
Keine Assistenz							

d.

	trifft gar nicht zu						trifft völlig zu
Assistenz zur Griffstabilität (permanent)							

e.

	trifft gar nicht zu						trifft völlig zu
Assistenz zur Griffstabilität (adaptiv)							

f.

	trifft gar nicht zu						trifft völlig zu
Assistenz zur Handposition							

VP-Nr.:

2. Mit der folgenden Assistenzvariante fand ich es schwierig, den Überblick zu behalten.

a.

Kombination Assistenz zur Handposition, Assistenz zur Griffstabilität (permanent)							
	trifft gar nicht zu						trifft völlig zu

b.

Kombination Assistenz zur Handposition, Assistenz zur Griffstabilität (adaptiv)							
	trifft gar nicht zu						trifft völlig zu

c.

Assistenz zur Griffstabilität (permanent)							
	trifft gar nicht zu						trifft völlig zu

d.

Assistenz zur Griffstabilität (adaptiv)							
	trifft gar nicht zu						trifft völlig zu

e.

Assistenz zur Handposition							
	trifft gar nicht zu						trifft völlig zu

3. In folgender Assistenzvariante war es einfach, die visuelle Information zu interpretieren.

a.

Kombination Assistenz zur Handposition, Assistenz zur Griffstabilität (permanent)							
	trifft gar nicht zu						trifft völlig zu

b.

Kombination Assistenz zur Handposition, Assistenz zur Griffstabilität (adaptiv)							
	trifft gar nicht zu						trifft völlig zu

c.

Assistenz zur Griffstabilität (permanent)							
	trifft gar nicht zu						trifft völlig zu

d.

Assistenz zur Griffstabilität (adaptiv)							
	trifft gar nicht zu						trifft völlig zu

e.

Assistenz zur Handposition							
	trifft gar nicht zu						trifft völlig zu

VP-Nr.:

4. Folgende Assistenzvariante war unnötig komplex.

a.

Kombination Assistenz zur Handposition, Assistenz zur Griffstabilität (permanent)							
	trifft gar nicht zu						trifft völlig zu

b.

Kombination Assistenz zur Handposition, Assistenz zur Griffstabilität (adaptiv)							
	trifft gar nicht zu						trifft völlig zu

c.

Assistenz zur Griffstabilität (permanent)							
	trifft gar nicht zu						trifft völlig zu

d.

Assistenz zur Griffstabilität (adaptiv)							
	trifft gar nicht zu						trifft völlig zu

e.

Assistenz zur Handposition							
	trifft gar nicht zu						trifft völlig zu

5. Ich kann mir vorstellen, dass die meisten Leute schnell lernen würden, mit der Assistenzvariante zu arbeiten.

a.

Kombination Assistenz zur Handposition, Assistenz zur Griffstabilität (permanent)							
	trifft gar nicht zu						trifft völlig zu

b.

Kombination Assistenz zur Handposition, Assistenz zur Griffstabilität (adaptiv)							
	trifft gar nicht zu						trifft völlig zu

c.

Assistenz zur Griffstabilität (permanent)							
	trifft gar nicht zu						trifft völlig zu

d.

Assistenz zur Griffstabilität (adaptiv)							
	trifft gar nicht zu						trifft völlig zu

e.

Assistenz zur Handposition							
	trifft gar nicht zu						trifft völlig zu

VP-Nr.:

6. Ich halte die Assistenzvariante für zu unvorhersehbar.
   a.

Kombination Assistenz zur Handposition, Assistenz zur Griffstabilität (permanent)							
	trifft gar nicht zu						trifft völlig zu

   b.

Kombination Assistenz zur Handposition, Assistenz zur Griffstabilität (adaptiv)							
	trifft gar nicht zu						trifft völlig zu

   c.

Assistenz zur Griffstabilität (permanent)							
	trifft gar nicht zu						trifft völlig zu

   d.

Assistenz zur Griffstabilität (adaptiv)							
	trifft gar nicht zu						trifft völlig zu

   e.

Assistenz zur Handposition							
	trifft gar nicht zu						trifft völlig zu

7. Ich fand es sehr mühsam, mit der Assistenzvariante zu arbeiten.
   a.

Kombination Assistenz zur Handposition, Assistenz zur Griffstabilität (permanent)							
	trifft gar nicht zu						trifft völlig zu

   b.

Kombination Assistenz zur Handposition, Assistenz zur Griffstabilität (adaptiv)							
	trifft gar nicht zu						trifft völlig zu

   c.

Assistenz zur Griffstabilität (permanent)							
	trifft gar nicht zu						trifft völlig zu

   d.

Assistenz zur Griffstabilität (adaptiv)							
	trifft gar nicht zu						trifft völlig zu

   e.

Assistenz zur Handposition							
	trifft gar nicht zu						trifft völlig zu

VP-Nr.:

8. Durch die Assistenz fühlte ich mich sehr sicher bei der Ausführung der Aufgabe.

a.

Kombination Assistenz zur Handposition, Assistenz zur Griffstabilität (permanent)							
	trifft gar nicht zu						trifft völlig zu

b.

Kombination Assistenz zur Handposition, Assistenz zur Griffstabilität (adaptiv)							
	trifft gar nicht zu						trifft völlig zu

c.

Assistenz zur Griffstabilität (permanent)							
	trifft gar nicht zu						trifft völlig zu

d.

Assistenz zur Griffstabilität (adaptiv)							
	trifft gar nicht zu						trifft völlig zu

e.

Assistenz zur Handposition							
	trifft gar nicht zu						trifft völlig zu

9. Die visuelle Darstellung änderte sich so schnell, dass es schwierig war, sie zu interpretieren.

a.

Assistenz zur Griffstabilität (permanent)							
	trifft gar nicht zu						trifft völlig zu

b.

Assistenz zur Griffstabilität (adaptiv)							
	trifft gar nicht zu						trifft völlig zu

c.

Assistenz zur Handposition							
	trifft gar nicht zu						trifft völlig zu

10. Bei folgender Assistenzvariante lieferten mir die Greifregionen wichtige Informationen, an denen ich mich orientierte.

a.

Assistenz zur Griffstabilität (permanent)							
	trifft gar nicht zu						trifft völlig zu

b.

Assistenz zur Griffstabilität (adaptiv)							
	trifft gar nicht zu						trifft völlig zu

c.

Assistenz zur Handposition							
	trifft gar nicht zu						trifft völlig zu

VP-Nr.:

11. Bei dieser Assistenzvariante kontrollierte ich meine Greifposition/Fingerposition durch das Schließen der Hand.

a.

Kombination Assistenz zur Handposition, Assistenz zur Griffstabilität (permanent)							
	trifft gar nicht zu						trifft völlig zu

b.

Kombination Assistenz zur Handposition, Assistenz zur Griffstabilität (adaptiv)							
	trifft gar nicht zu						trifft völlig zu

c.

Assistenz zur Griffstabilität (permanent)							
	trifft gar nicht zu						trifft völlig zu

d.

Assistenz zur Griffstabilität (adaptiv)							
	trifft gar nicht zu						trifft völlig zu

e.

Assistenz zur Handposition							
	trifft gar nicht zu						trifft völlig zu

12. Bei der Assistenz zur **Griffstabilität** gab es Situationen, in denen ich nicht wusste, was ich ändern musste, um den Griff zu verbessern.

trifft gar nicht zu						trifft völlig zu

13. Die Assistenz zur **Griffstabilität** war vollkommen ausreichend, um einen stabilen Griff zu finden.

trifft gar nicht zu						trifft völlig zu

14. Es reichte aus, wenn die Assistenz zur **Griffstabilität** nur beim Schließen der Hand erschien.

trifft gar nicht zu						trifft völlig zu

15. Bei der adaptiven Assistenz zur **Griffstabilität** achtete ich beim Schließen der Finger nur auf die angezeigte Farbe, nicht auf die Fingerposition.

trifft gar nicht zu						trifft völlig zu

16. Die permanente Assistenz zur **Griffstabilität** lenkte mich ab.

trifft gar nicht zu						trifft völlig zu

VP-Nr.:

17. Bei der Assistenz zur **Handposition** fand ich es schwierig, die notwendige Handbewegung nachzuvollziehen.

☐	☐	☐	☐	☐	☐	☐

trifft gar nicht zu — trifft völlig zu

18. Bei der Assistenz zur **Handposition** gab es Situationen, in denen ich nicht wusste, was ich ändern musste, um den Griff zu verbessern.

☐	☐	☐	☐	☐	☐	☐

trifft gar nicht zu — trifft völlig zu

19. Bei der Assistenz zur **Handposition** war mir nicht klar, wie ich die Hand bewegen muss, um die gewünschte Position zu erreichen.

☐	☐	☐	☐	☐	☐	☐

trifft gar nicht zu — trifft völlig zu

20. Die Assistenz zur **Handposition** reichte völlig aus, um einen Griff zu finden, sodass ich die Greifregionen nicht weiter beachtete.

☐	☐	☐	☐	☐	☐	☐

trifft gar nicht zu — trifft völlig zu

21. Bei der Assistenz zur **Handposition** konnte ich nachvollziehen, warum eine von mir gewählte Handposition nicht zu einem stabilen Griff führte.

☐	☐	☐	☐	☐	☐	☐

trifft gar nicht zu — trifft völlig zu

22. Es gab Momente, in denen ich die Assistenz zur **Handposition** nicht nachvollziehen konnte, weil der Pfeil von der Hand überdeckt wurde.

☐	☐	☐	☐	☐	☐	☐

trifft gar nicht zu — trifft völlig zu

23. Die Informationsdarstellung war auch dann übersichtlich, wenn sowohl die Assistenz zur **Handposition** als auch die Assistenz zur **Griffstabilität** gezeigt wurden.

☐	☐	☐	☐	☐	☐	☐

trifft gar nicht zu — trifft völlig zu

VP-Nr.:

24. Selbst bei „gespiegelter“ Hand war das Greifen mit folgender Assistenz einfach.

a.

Kombination Assistenz zur Handposition, Assistenz zur Griffstabilität (permanent)							
	trifft gar nicht zu						trifft völlig zu

b.

Kombination Assistenz zur Handposition, Assistenz zur Griffstabilität (adaptiv)							
	trifft gar nicht zu						trifft völlig zu

c.

Assistenz zur Griffstabilität (permanent)							
	trifft gar nicht zu						trifft völlig zu

d.

Assistenz zur Griffstabilität (adaptiv)							
	trifft gar nicht zu						trifft völlig zu

e.

Assistenz zur Handposition							
	trifft gar nicht zu						trifft völlig zu

25. Insgesamt finde ich folgende Assistenzvariante am hilfreichsten:

a.

Kombination Assistenz zur Handposition, Assistenz zur Griffstabilität (permanent)							
	trifft gar nicht zu						trifft völlig zu

b.

Kombination Assistenz zur Handposition, Assistenz zur Griffstabilität (adaptiv)							
	trifft gar nicht zu						trifft völlig zu

c.

Assistenz zur Griffstabilität (permanent)							
	trifft gar nicht zu						trifft völlig zu

d.

Assistenz zur Griffstabilität (adaptiv)							
	trifft gar nicht zu						trifft völlig zu

e.

Assistenz zur Handposition							
	trifft gar nicht zu						trifft völlig zu

26. Sonstige Bemerkungen zu den Assistenzvarianten/Vorschläge zur weiteren Optimierung:

____

____

VP-Nr.:

27. Die **Position der Hand** hat mich bei der Interaktion stark gestört.

a.

„natürliche" Position							
	trifft gar nicht zu						trifft völlig zu

b.

„gespiegelte" Position							
	trifft gar nicht zu						trifft völlig zu

28. Nach einer kurzen Eingewöhnung habe ich nicht mehr über die Position der Hand nachgedacht.

a.

„natürliche" Position							
	trifft gar nicht zu						trifft völlig zu

b.

„gespiegelte" Position							
	trifft gar nicht zu						trifft völlig zu

VP-Nr.:

29. Während der Interaktion wurde ich durch den folgenden Faktor eingeschränkt.

a. Head-Mounted Display

	trifft gar nicht zu						trifft völlig zu
3D-Sicht							
Auflösung des Displays							
Gewicht							
Bewegungseinschränkung							

b. Eingabestation HUG

	trifft gar nicht zu						trifft völlig zu
Schwergängigkeit							
Gewicht							
Bewegungseinschränkung							

c.

	trifft gar nicht zu						trifft völlig zu
Lärm/Geräusche aus der Umgebung							

d.

	trifft gar nicht zu						trifft völlig zu
Visuelle Reize aus der Umgebung							

e.

	trifft gar nicht zu						trifft völlig zu
Assistenz zur Griffstabilität (adaptiv)							

f.

	trifft gar nicht zu						trifft völlig zu
Assistenz zur Handposition							

Sonstige Faktoren: ____

____

A.3. Study 3

A.3.1. Questionnaire "TLX" and Naturalism

ParticipantNr: _____

Age: _____________ Years

Sex: m w

I am right / left handed (please underline)

Professional Background ______________________________

1. How many hours do you use CAD programs per week (on average)?

☐ **hours**

2. How many hours do you play computer games per week (on average)?

☐ **hours**

3. I worked with the system previous to the user study

yes ☐ no ☐

If yes, how often?______________________________

Comments (filled by investigator):

____

____

____

Order of conditions:

1: teleoperation mode
2: semi-autonomous mode (grasp when far)
3: semi-autonomous mode (grasp when close)

Short-questionnaire Usability-Study

ParticipantNr: _____ SpaceMouse-mode:

NORMAL

Teleoperation – mode:

1. How high was your sense of being physically present in the virtual environment (= sense of presence; 0% = no presence to 100% = sense of presence like in the real world).

[ ] **%**

2.1. Mental demand	How much mental demand was needed to intake information and process it? Was the task easy or complex?
Very low/ easy	very high/ complex
2.2 Physical demand	How much physical demand was needed? Was the task easy or exhausting?
Very low/ easy	very high/ exhausting

2.3 How much did you feel restricted during execution of the task?	Not at all						Very much
2.4. How naturally could you control the virtual robot?	Not at all						Very much
2.5 How naturally did the robot move?	Not at all						Very much
2.6 The behavior of the robot matched my expectations	Not at all						Very much
2.7 The positions selected for grasping were human like	Not at all						Very much

NORMAL

Semi-autonomous – mode (grasp when close):

1. How high was your sense of being physically present in the virtual environment (= sense of presence; 0% = no presence to 100% = sense of presence like in the real world).

[ ] %

2.1. Mental demand	How much mental demand was needed to intake information and process it? Was the task easy or complex?
Very low/ easy	very high/ complex
2.2 Physical demand	How much physical demand was needed? Was the task easy or exhausting?
Very low/ easy	very high/ exhausting

2.3 How much did you feel restricted during execution of the task?							
	Not at all						Very much
2.4. How naturally could you control the virtual robot?							
	Not at all						Very much
2.5 How naturally did the robot move?							
	Not at all						Very much
2.6 The behavior of the robot matched my expectations							
	Not at all						Very much
2.7 The positions selected for grasping were human like							
	Not at all						Very much

NORMAL

Semi-autonomous – mode (grasp when far):

1. How high was your sense of being physically present in the virtual environment (= sense of presence; 0% = no presence to 100% = sense of presence like in the real world).

[ ] **%**

2.1. Mental demand	How much mental demand was needed to intake information and process it? Was the task easy or complex?
Very low/ easy	very high/ complex
2.2 Physical demand	How much physical demand was needed? Was the task easy or exhausting?
Very low/ easy	very high/ exhausting

	Not at all						Very much
2.3 How much did you feel restricted during execution of the task?							
2.4. How naturally could you control the virtual robot?							
2.5 How naturally did the robot move?							
2.6 The behavior of the robot matched my expectations							
2.7 The positions selected for grasping were human like							

A.3.2. Questionnaire "Attrak-Diff"

Questionnaire

ParticipantNr: _____ SpaceMouse-mode: **normal**

E.1 In following condition, it was easy for me to grasp the object

a) teleoperation							
	Not at all						Very much
b) semi-autonomous (grasp when close)							
	Not at all						Very much
c) semi-autonomous (grasp when far)							
	Not at all						Very much

E.2 In following condition, the robot moved in a predictable way

a) teleoperation							
	Not at all						Very much
b) semi-autonomous (grasp when close)							
	Not at all						Very much
c) semi-autonomous (grasp when far)							
	Not at all						Very much

E.3 In following condition, I understood when I could grasp the object

a) teleoperation							
	Not at all						Very much
b) semi-autonomous (grasp when close)							
	Not at all						Very much
c) semi-autonomous (grasp when far)							
	Not at all						Very much

E.4 I had the impression that there was a time delay between my commands and the movement of the robot

a) teleoperation							
	Not at all						Very much
b) semi-autonomous (grasp when close)							
	Not at all						Very much
c) semi-autonomous (grasp when far)							
	Not at all						Very much

E5. Rate the usefulness of the modes of operation to grasp an object

a) teleoperation							
	Not at all						Very much
b) semi-autonomous (grasp when close)							
	Not at all						Very much
c) semi-autonomous (grasp when far)							
	Not at all						Very much

Other comments:

____

____

____

Questionnaire

ParticipantNr: _____ SpaceMouse-mode: **distorted**

E.1 In following condition, it was easy for me to grasp the object

	Not at all						Very much
a) teleoperation							
b) semi-autonomous (grasp when close)							
c) semi-autonomous (grasp when far)							

E.2 In following condition, the robot moved in a predictable way

	Not at all						Very much
a) teleoperation							
b) semi-autonomous (grasp when close)							
c) semi-autonomous (grasp when far)							

E.3 In following condition, I understood when I could grasp the object

	Not at all						Very much
a) teleoperation							
b) semi-autonomous (grasp when close)							
c) semi-autonomous (grasp when far)							

E.4 I had the impression that there was a time delay between my commands and the movement of the robot

	Not at all						Very much
a) teleoperation							
b) semi-autonomous (grasp when close)							
c) semi-autonomous (grasp when far)							

E5. Rate the usefulness of the modes of operation to grasp an object

	Not at all						Very much
a) teleoperation							
b) semi-autonomous (grasp when close)							
c) semi-autonomous (grasp when far)							

Other comments:

____

____

____

List of Figures

List of Tables

Bibliography

[1] 3DConnexion. *SpaceMouse*. Online; accessed 5-January-2015.
URL: http://www.3dconnexion.de/.

[2] D. Aarno, S. Ekvall, and D. Kragić. "Adaptive Virtual Fixtures for Machine-Assisted Teleoperation Tasks". In: *Proc. IEEE Int. Conf. on Robotics and Automation (ICRA)*. Apr. 2005, pp. 897–903.
DOI: 10.1109/ROBOT.2005.1570231.

[3] J.J. Abbott, G.D. Hager, and A.M. Okamura. "Steady-hand teleoperation with virtual fixtures". In: *Proc. IEEE Int. Workshop on Robot and Human Interactive Communication (ROMAN)*. Oct. 2003, pp. 145–151.
DOI: 10.1109/ROMAN.2003.1251824.

[4] J.J. Abbott, P. Marayong, and A.M. Okamura. "Haptic Virtual Fixtures for Robot-Assisted Manipulation". In: *Robotics Research*. Ed. by S. Thrun, R. Brooks, and H. Durrant-Whyte. Vol. 28. Springer Tracts in Advanced Robotics. Springer Berlin Heidelberg, 2007, pp. 49–64. ISBN: 978-3-540-48110-2.
DOI: 10.1007/978-3-540-48113-3_5.

[5] A. Adams, P. Rani, and N. Sarkar. *Mixed Initiative Interaction and Robotic Systems*. 2004.
URL: http://www.vuse.vanderbilt.edu/~adamsja/Papers/WS1604JAdams.pdf.

[6] D. Alba, M. Armada, and R. Ponticelli. "An Introductory Revision to Humanoid Robot Hands". In: *Climbing and Walking Robots*. Springer Berlin Heidelberg, 2005, pp. 701–712. ISBN: 978-3-540-22992-6.
DOI: 10.1007/3-540-29461-9_69.

[7] R. Anderson. "Autonomous, teleoperated, and shared control of robot systems". In: *Proc. IEEE Int. Conf. on Robotics and Automation (ICRA)*. Vol. 3. Apr. 1996, pp. 2025–2032.
DOI: 10.1109/ROBOT.1996.506169.

[8] R. Aracil, M. Buss, S. Cobos, M. Ferre, S. Hirche, M. Kuschel, and A. Peer. "The Human Role in Telerobotics". In: *Advances in Telerobotics*. Ed. by M. Ferre, M. Buss, R. Aracil, C. Melchiorri, and C. Balaguer. Vol. 31. Springer Tracts in Advanced Robotics. Springer Berlin Heidelberg, 2007.
DOI: 10.1007/978-3-540-71364-7_2.

[9] K.J. Åström and B. Wittenmark. *Adaptive Control*. Dover Books on Electrical Engineering Series. Dover Publications, 2008. ISBN: 9780486462783.
URL: http://books.google.de/books?id=L0m_CR-IK24C.

[10] P.G. Backes and K.S. Tso. "UMI: an interactive supervisory and shared control system for telerobotics". In: *Proc. IEEE Int. Conf. on Robotics and Automation (ICRA)*. Vol. 2. May 1990, pp. 1096–1101.
DOI: 10.1109/ROBOT.1990.126141.

[11] L. Bainbridge. "Ironies of automation". In: *Automatica* 19 (6 Nov. 1983), pp. 775–779.
URL: https://www.ise.ncsu.edu/nsf_itr/794B/papers/Bainbridge_1983_Automatica.pdf.

[12] C.B. Barber, D.P. Dobkin, and H. Huhdanpaa. "The quickhull algorithm for convex hulls". In: *ACM Transactions on Mathematical Software* 22.4 (1996), pp. 469–483.
URL: http://www.cise.ufl.edu/~ungor/courses/fall06/papers/QuickHull.pdf.

[13] D. Barber, L. Davis, D. Kemper, P. Smith, and D. Nicholson. "Collaborative human robot interactions in combined arms operations". In: *Int. Symp. on Collaborative Technologies and Systems (CTS)*. May 2007, pp. 88–92.
DOI: 10.1109/CTS.2007.4621742.

[14] L. Basañez and R. Suárez. "Teleoperation". In: *Springer Handbook of Automation*. Ed. by Shimon Y. Nof. Springer Berlin Heidelberg, 2009, pp. 449–468. ISBN: 978-3-540-78830-0.
DOI: 10.1007/978-3-540-78831-7_27.

[15] Z. Beauchamp and J. Savulescu. "Robot Guardians: Teleoperated Combat Vehicles in Humanitarian Military Intervention". In: *In Killing by Remote Control: The Ethics of an Unmanned Military*. Ed. by B.J. Strawser and J. McMahan. New York: Oxford University Press, 2013.
DOI: 10.1093/acprof:oso/9780199926121.003.0006.

[16] J.M. Beer, A.D. Fisk, and W.A. Rogers. "Toward a framework for levels of robot autonomy in human-robot interaction". In: *Journal of Human-Robot Interaction* 3.2 (2014), pp. 74–99.
URL: http://dx.doi.org/10.5898/JHRI.3.2.Beer.

[17] A.K. Bejczy and Won S.K. "Predictive displays and shared compliance control for time-delayed telemanipulation". In: *Proc. IEEE/RSJ Int. Conf. on Intelligent Robots and Systems (IROS)*. Vol. 1. July 1990, pp. 407–412.
DOI: 10.1109/IROS.1990.262418.

[18] D. Berenson, S.S. Srinivasa, and J. Kuffner. "Addressing Pose Uncertainty in Manipulation Planning Using Task Space Regions". In: *Proc. IEEE/RSJ Int. Conf. on Intelligent Robots and Systems (IROS)*. 2009, pp. 1419–1425.
DOI: 10.1109/IROS.2009.5354833.

[19] R. Berka, S.M. Goza, E. Potter, C. Edmondson, and N. De La Pena. "Tele-Operation Control of Robonaut on the International Space Station". In: *IEEE Aerospace*. 2012.
URL: http://dx.doi.org/10.2514/6.2012-2427.

[20] L. Biagiotti, F. Lotti, C. Melchiorri, and G. Vassura. "How far is the human hand? A Review on Anthropomorphic Robotic End-Effectors". In: *Internal Report - University of Bologna* (2004).
URL: http://www-lar.deis.unibo.it/woda/data/deis-lar-publications/3cbd.Document.pdf.

[21] A. Bicchi. "Hands for Dexterous Manipulation and Robust Grasping: A Difficult Road Towards Simplicity". In: *IEEE Trans. on Robotics and Automation* 16.6 (2000), pp. 652–662.
DOI: 10.1109/70.897777.

[22] A. Bicchi and V. Kumar. "Robotic Grasping and Contact: A Review". In: *Proc. IEEE Int. Conf. on Robotics and Automation (ICRA)*. 2000, pp. 348–352.
DOI: 10.1109/ROBOT.2000.844081.

[23] P. Birkenkampf, D. Leidner, and C. Borst. "A Knowledge-Driven Shared Autonomy Human-Robot Interface for Tablet Computers". In: *Proc. IEEE-RAS Int. Conf. on Humanoid Robots*. Nov. 2014.
DOI: 10.1109/HUMANOIDS.2014.7041352.

[24] H. Boessenkool, D.A. Abbink, C.J.M. Heemskerk, F.C.T. van der Helm, and J.G.W. Wildenbeest. "A Task-Specific Analysis of the Benefit of Haptic Shared Control During Telemanipulation". In: *IEEE Trans. on Haptics* 6.1 (2013), pp. 2–12.
DOI: 10.1109/TOH.2012.22.

[25] J. Bohg, A. Morales, T. Asfour, and D. Kragic. "Data-Driven Grasp Synthesis - A Survey". In: *IEEE Trans. on Robotics* 30.2 (2014), pp. 289–309.
DOI: 10.1109/TRO.2013.2289018.

[26] J. Bohren, C. Papazov, D. Burschka, K. Krieger, S. Parusel, S. Haddadin, W. Shepherdson, G. Hager, and L. Whitcomb. "A Pilot Study in Vision-Based Augmented Telemanipulation for Remote Assembly Over High-Latency Networks". In: *Proc. IEEE Int. Conf. on Robotics and Automation (ICRA)*. 2013, pp. 3631–3638.
DOI: 10.1109/ICRA.2013.6631087.

[27] C. Borst, C. Ott, T. Wimböck, B. Brunner, F. Zacharias, B. Bauml, U. Hillenbrand, S. Haddadin, A. Albu-Schäffer, and G. Hirzinger. "A humanoid upper body system for two-handed manipulation". In: *Proc. IEEE Int. Conf. on Robotics and Automation (ICRA)*. 2007, pp. 2766–2767.
DOI: 10.1109/ROBOT.2007.363886.

[28] Ch. Borst, M. Fischer, and G. Hirzinger. "A Fast and Robust Grasp Planner for Arbitrary 3D Objects". In: *Proc. IEEE Int. Conf. on Robotics and Automation (ICRA)*. 1999, pp. 1890–1896.
DOI: 10.1109/ROBOT.1999.770384.

[29] Ch. Borst, M. Fischer, and G. Hirzinger. "Calculating hand configurations for precision and pinch grasps". In: *Proc. IEEE/RSJ Int. Conf. on Intelligent Robots and Systems (IROS)*. 2002, pp. 1553–1559.
DOI: 10.1109/IRDS.2002.1043976.

[30] Ch. Borst, M. Fischer, and G. Hirzinger. "Grasp Planning: How to Choose a Suitable Task Wrench Space". In: *Proc. IEEE Int. Conf. on Robotics and Automation (ICRA)*. 2004, pp. 319–325.
DOI: 10.1109/ROBOT.2004.1307170.

[31] S.A. Bowyer, B.L. Davies, and F. Rodriguez y Baena. "Active Constraints/Virtual Fixtures: A Survey". In: *IEEE Trans. on Robotics* 30.1 (Feb. 2014), pp. 138–157.
DOI: 10.1109/TRO.2013.2283410.

[32] S.P. Boyd and B. Wegbreit. "Fast Computation of Optimal Contact Forces". In: *IEEE Trans. on Robotics* 23.6 (2007), pp. 1117–1132.
DOI: 10.1109/TRO.2007.910774.

[33] M. Brucker, S. Leonard, T. Bodenmüller, and G. D. Hager. "Sequential scene parsing using range and intensity information". In: *Proc. IEEE Int. Conf. on Robotics and Automation (ICRA)*. 2012, pp. 5417–5424.
DOI: 10.1109/ICRA.2012.6224978.

[34] D.J. Bruemmer, J.L. Marble, Donald D. Dudenhoeffer, M.O. Anderson, and M.D. Mckay. "Mixed-initiative control for remote characterization of hazardous environments". In: *Proc. Annual Hawaii Int. Conf. on System Sciences*. Jan. 2003.
DOI: 10.1109/HICSS.2003.1174289.

[35] B. Brunner, G. Hirzinger, K. Landzettel, and J. Heindl. "Multisensory shared autonomy and tele-sensor-programming-Key issues in the space robot technology experiment ROTEX". In: *Proc. IEEE/RSJ Int. Conf. on Intelligent Robots and Systems (IROS)*. Vol. 3. July 1993, pp. 2123–2139.
DOI: 10.1109/IROS.1993.583924.

[36] M. Buss, H. Hashimoto, and J.B. Moore. "Dextrous Hand Grasping Force Optimization". In: *IEEE Trans. on Robotics and Automation* 12.3 (1996), pp. 406–418.
DOI: 10.1109/70.499823.

[37] M. Buss, A. Peer, T. Schauss, N. Stefanov, U. Unterhinninghofen, S. Behrendt, J. Leupold, M. Durkovic, and M. Sarkis. "Development of a Multi-modal Multi-user Telepresence and Teleaction System". In: *Int. J. Robotics Research* 29.10 (2010), pp. 1298–1316.
DOI: 10.1177/0278364909351756.

[38] J. Butterfass, M. Grebenstein, H. Liu, and G. Hirzinger. "DLR Hand II: Next Generation of a Dextrous Robot Hand". In: *Proc. IEEE Int. Conf. on Robotics and Automation (ICRA)*. 2001, pp. 109–114.
DOI: 10.1109/ROBOT.2001.932538.

[39] K.-E. Bystrom, W. Barfield, and C. Hendrix. "A Conceptual Model of the Sense of Presence in Virtual Environments". In: *Presence: Teleoper. and Virtual Environ.* 8.2 (Apr. 1999), pp. 241–244.
DOI: 10.1162/105474699566107.

[40] U. Castiello. "The neuroscience of grasping". In: *Nature Reviews Neuroscience* 6.9 (Sept. 2005), pp. 726–736.
DOI: 10.1038/nrn1744.

[41] K. Charusta, R. Krug, T. Stoyanov, D. Dimitrov, and B. Iliev. "Generation of Independent Contact Regions on Objects Reconstructed from Noisy Real-World Range Data". In: *Proc. IEEE Int. Conf. on Robotics and Automation (ICRA)*. 2012, pp. 1338–1344.
DOI: 10.1109/ICRA.2012.6225046.

[42] J.Y.C. Chen, E.C. Haas, and M.J. Barnes. "Human Performance Issues and User Interface Design for Teleoperated Robots". In: *IEEE Trans. on Systems, Man, and Cybernetics; Part C: Applications and Reviews* 37.6 (Nov. 2007), pp. 1231–1245.
DOI: 10.1109/TSMCC.2007.905819.

[43] M. Ciocarlie, K. Hsiao, A. Leeper, and D. Gossow. "Mobile manipulation through an assistive home robot". In: *Proc. IEEE/RSJ Int. Conf. on Intelligent Robots and Systems (IROS)*. 2012, pp. 5313–5320.
DOI: 10.1109/IROS.2012.6385907.

[44] M.T. Ciocarlie and P.K. Allen. "Hand Posture Subspaces for Dexterous Robotic Grasping". In: *Int. J. Robotics Research* 28.7 (2009), pp. 851–867.
DOI: 10.1177/0278364909105606.

[45] L. Colasanto, R. Suarez, and J. Rosell. "Hybrid Mapping for the Assistance of Teleoperated Grasping Tasks". In: *IEEE Trans. on Systems, Man, and Cybernetics* 43.2 (Mar. 2013), pp. 390–401.
DOI: 10.1109/TSMCA.2012.2195309.

[46] B.-A. Dang-Vu, M.A. Roa, and C. Borst. "Extended independent contact regions for grasping applications". In: *Proc. IEEE/RSJ Int. Conf. on Intelligent Robots and Systems (IROS)*. Nov. 2013, pp. 3527–3534.
DOI: 10.1109/IROS.2013.6696859.

[47] M. Desai, K.M. Tsui, H.A Yanco, and C. Uhlik. "Essential features of telepresence robots". In: *IEEE Conf. on Technologies for Practical Robot Applications (TePRA)*. Apr. 2011, pp. 15–20.
DOI: 10.1109/TEPRA.2011.5753474.

[48] P. Desbats, F. Geffard, G. Piolain, and A. Coudray. "Force-feedback teleoperation of an industrial robot in a nuclear spent fuel reprocessing plant". In: *Industrial Robot: An International Journal* 33.3 (2006), pp. 178–186.
URL: http://www-ist.cea.fr/publicea/exl-doc/200600000132.pdf.

[49] Rosen Diankov. "Automated Construction of Robotic Manipulation Programs". PhD thesis. Carnegie Mellon University, Robotics Institute, 2010.
URL: http://repository.cmu.edu/dissertations/32/.

[50] Force Dimension. *Sigma.7*. Online; accessed 5-January-2015. Route de Saint-Cergue 295, CH-1260 Nyon, Switzerland, 2014.
URL: http://www.forcedimension.com/products/sigma-7/overview.

[51] L. Dipietro, A.M. Sabatini, and P. Dario. "A Survey of Glove-Based Systems and Their Applications". In: *IEEE Trans. on Systems, Man, and Cybernetics; Part C: Applications and Reviews* 38.4 (July 2008), pp. 461–482.
DOI: 10.1109/TSMCC.2008.923862.

[52] J.V. Draper. "Teleoperators for advanced manufacturing: Applications and human factors challenges". In: *Int. Journal of Human Factors in Manufacturing* 5.1 (1995), pp. 53–85.
DOI: 10.1002/hfm.4530050105.

[53] J.V. Draper, D.B. Kaber, and J.M. Usher. "Telepresence". In: *Human Factors* 40.3 (Sept. 1998).

[54] B. Drost, M. Ulrich, N. Navab, and S. Ilic. "Model globally, match locally: Efficient and robust 3D object recognition". In: *CVPR*. 2010, pp. 998–1005.
DOI: 10.1109/CVPR.2010.5540108.

[55] R.V. Dubey, S.E. Everett, N. Pernalete, and K. A Manocha. "Teleoperation assistance through variable velocity mapping". In: *IEEE Trans. on Robotics and Automation* 17.5 (Oct. 2001), pp. 761–766.
DOI: 10.1109/70.964674.

[56] S. El Khoury, M. Li, and A. Billard. "Bridging the Gap: One shot grasp synthesis approach". In: *Proc. IEEE/RSJ Int. Conf. on Intelligent Robots and Systems (IROS)*. 2012, pp. 2027–2034.
DOI: 10.1109/IROS.2012.6385886.

[57] M.R. Endsley and D.B. Kaber. "Level of automation effects on performance, situation awareness and workload in a dynamic control task". In: *Ergonomics* 42.3 (1999), pp. 462–492.
URL: http://people.engr.ncsu.edu/dbkaber/papers/Endsley_Kaber_Ergo_99.pdf.

[58] M.R. Endsley and E.O. Kiris. "The Out-of-the-Loop Performance Problem and Level of Control in Automation". In: *Human Factors: J. of Human Factors and Ergonomics Society* 37.2 (1995), pp. 381–394.
DOI: 10.1518/001872095779064555.

[59] E. Ettelt, R. Furtwängler, U.D. Hanebeck, and G. Schmidt. "Design issues of a semi-autonomous robotic assistant for the health care environment". In: *Journal of Intelligent and Robotic Systems* 22.3-4 (1998), pp. 191–209.
DOI: 10.1023/A:1008082024638.

[60] M. Fallon, S. Kuindersma, S. Karumanchi, M. Antone, T. Schneider, H. Dai, C.P. D'Arpino, R. Deits, M. DiCicco, D. Fourie, T. Koolen, P. Marion, M. Posa, A. Valenzuela, K.-T. Yu, J. Shah, K. Iagnemma, R. Tedrake, and S. Teller. "An Architecture for Online Affordance-based Perception and Whole-body Planning". In: *J. Field Robotics* (2014).
DOI: 10.1002/rob.21546.

[61] C. Ferrari and J. Canny. "Planning Optimal Grasps". In: *Proc. IEEE Int. Conf. on Robotics and Automation (ICRA)*. 1992, pp. 2290–2295.
DOI: 10.1109/ROBOT.1992.219918.

[62] M. Fischer, P. van der Smagt, and G. Hirzinger. "Learning techniques in a dataglove based telemanipulation system for the DLR hand". In: *Proc. IEEE Int. Conf. on Robotics and Automation (ICRA)*. Vol. 2. May 1998, pp. 1603–1608.
DOI: 10.1109/ROBOT.1998.677377.

[63] T.W. Fong, C. Thorpe, and C. Baur. "Collaboration, Dialogue, and Human-Robot Interaction". In: *Proc. Int. Symp. of Robotics Research*. Springer-Verlag, Nov. 2001.
DOI: 10.1007/3-540-36460-9_17.

[64] J. Funda and R.P. Paul. "Teleprogramming: overcoming communication delays in remote manipulation". In: *Proc. Int. Conf. on Systems, Man and Cybernetics*. Nov. 1990, pp. 873–875.
DOI: 10.1109/ICSMC.1990.142248.

[65] Willow Garage. *PR 2, personal robot*. Online; accessed 5-January-2015.
URL: https://www.willowgarage.com/pages/pr2/overview.

[66] L. Geiger, M. Popp, B. Färber, J. Artigas, and P. Kremer. "The Influence of Telemanipulation-Systems on Fine Motor Performance". In: *Proc. Int. Conf. on Advances in Computer-Human Interactions (ACHI)*. Feb. 2010, pp. 44–49.
DOI: 10.1109/ACHI.2010.46.

[67] T.L. Gibo, AJ. Bastian, and AM. Okamura. "Grip Force Control during Virtual Object Interaction: Effect of Force Feedback, Accuracy Demands, and Training". In: *IEEE Trans. on Haptics* 7.1 (Jan. 2014), pp. 37–47.
DOI: 10.1109/TOH.2013.60.

[68] M. Gienger, M. Toussaint, and C. Goerick. "Task maps in humanoid robot manipulation". In: *Proc. IEEE/RSJ Int. Conf. on Intelligent Robots and Systems (IROS)*. Sept. 2008, pp. 2758–2764.
DOI: 10.1109/IROS.2008.4651027.

[69] G. Gioioso, G. Salvietti, M. Malvezzi, and D. Prattichizzo. "Mapping Synergies From Human to Robotic Hands With Dissimilar Kinematics: An Approach in the Object Domain". In: *IEEE Trans. on Robotics* 29.4 (Aug. 2013), pp. 825–837.
DOI: 10.1109/TRO.2013.2252251.

[70] C. Goldfeder, P.K. Allen, C. Lackner, and R. Pelossof. "Grasp Planning via Decomposition Trees". In: *Proc. IEEE Int. Conf. on Robotics and Automation (ICRA)*. 2007, pp. 4679–4684.
DOI: 10.1109/ROBOT.2007.364200.

[71] C. Goldfeder, M. Ciocarlei, H. Dang, and P.K. Allen. "The Columbia Grasp Database". In: *Proc. IEEE Int. Conf. on Robotics and Automation (ICRA)*. 2009, pp. 1710–1716. DOI: 10.1109/ROBOT.2009.5152709.

[72] M.A. Goodrich, J.W. Crandall, and E. Barakova. "Teleoperation and Beyond for Assistive Humanoid Robots". In: *Reviews of Human Factors and Ergonomics* 9.1 (2013), pp. 175–226. DOI: 10.1177/1557234X13502463.

[73] W Griffin, W Provancher, and M Cutkosky. "Feedback Strategies for Telemanipulation with Shared Control of Object Handling Forces". In: *Presence: Teleoper. and Virtual Environ.* 14.6 (Dec. 2005), pp. 720–731. URL: http://www.mitpressjournals.org/doi/pdf/10.1162/105474605775196634.

[74] W.B. Griffin, R.P. Findley, M.L. Turner, and M.R. Cutkosky. "Calibration and mapping of a human hand for dexterous telemanipulation". In: *ASME IMECE Symp. on Haptic Interfaces for Virtual Environments and Teleoperator Systems*. 2000, pp. 1–8. URL: http://www-cdr.stanford.edu/dml/publications/griffin_asme00.pdf.

[75] L. Han, Y.S. Guan, Z.X. Li, Q. Shi, and J.C. Trinkle. "Dextrous Manipulation with Rolling Contacts". In: *Proc. IEEE Int. Conf. on Robotics and Automation (ICRA)*. 1997, pp. 992–997. DOI: 10.1109/ROBOT.1997.614264.

[76] L. Han, J.C. Trinkle, and Z.X. Li. "Grasp Analysis as Linear Matrix Inequality Problems". In: *IEEE Trans. on Robotics and Automation* 16.6 (2000), pp. 663–674. DOI: 10.1109/70.897778.

[77] B. Hannaford and L. Wood. "Performance Evaluation of a 6 Axis High Fidelity Generalized Force Reflecting Teleoperator". In: *Proc. JPL/NASA Conf. on Space Telerobotics*. Jan. 1989.

[78] K. Harada, K. Kaneko, and F. Kanehiro. "Fast Grasp Planning for Hand/Arm Systems Based on Convex Model". In: *Proc. IEEE Int. Conf. on Robotics and Automation (ICRA)*. 2008, pp. 1162–1168. DOI: 10.1109/ROBOT.2008.4543361.

[79] S. Hart and L. Staveland. "Development of NASA-TLX (Task Load Index): Results of empirical and theoretical research". In: *Human Mental Workload* (1988).

[80] M. Hassenzahl, M. Burmester, and F. Koller. "AttrakDiff: Ein Fragebogen zur Messung wahrgenommener hedonischer und pragmatischer Qualität". In: *Mensch und Computer 2003: Interaktion in Bewegung*. (2003), pp. 187–196.

[81] K. Hauser. "Recognition, prediction, and planning for assisted teleoperation of freeform tasks". In: *Autonomous Robots* 35.4 (2013), pp. 241–254. ISSN: 0929-5593. DOI: 10.1007/s10514-013-9350-3.

[82] Georgia Institute of Technology Healthcare Robotics Labs. *PR2 playpen*. URL: ros.org/wiki/pr2_playpen.

[83] S. Henderson and S. Feiner. "Exploring the Benefits of Augmented Reality Documentation for Maintenance and Repair". In: *IEEE Trans. on Visualization and Computer Graphics* 17.10 (Oct. 2011), pp. 1355–1368. DOI: 10.1109/TVCG.2010.245.

[84] P. Heo, G. Gu, S.-J. Lee, K. Rhee, and J. Kim. "Current hand exoskeleton technologies for rehabilitation and assistive engineering". In: *International Journal of Precision Engineering and Manufacturing* 13.5 (2012), pp. 807–824.
DOI: 10.1007/s12541-012-0107-2.

[85] S. Hirche. "Haptic Telepresence in Packet Switched Communication Networks". Dissertation. München: Technische Universität München, 2005.
URL: http://nbn-resolving.de/urn/resolver.pl?urn:nbn:de:bvb:91-diss20050919-2000268707.

[86] H. Hirschmüller. "Stereo Processing by Semi-Global Matching and Mutual Information". In: *IEEE Trans. on Pattern Analysis and Machine Intelligence* 30.2 (2008), pp. 328–341.
DOI: 10.1109/TPAMI.2007.1166.

[87] G. Hirzinger, B. Brunner, J. Dietrich, and J. Heindl. "ROTEX - the first remotely controlled robot in space". In: *Proc. IEEE Int. Conf. on Robotics and Automation (ICRA)*. May 1994, 2604–2611 vol.3.
DOI: 10.1109/ROBOT.1994.351121.

[88] G. Hirzinger, N. Sporer, A. Albu-Schäffer, M. Hähnle, R. Krenn, A. Pascucci, and M. Schedl. "DLR's torque-controlled light weight robot III – are we reaching the technological limits now?" In: *Proc. IEEE Int. Conf. on Robotics and Automation (ICRA)*. Vol. 2. 2002, pp. 1710–1716.
DOI: 10.1109/ROBOT.2002.1014788.

[89] L.R. Hochberg, D. Bacher, B. Jarosiewicz, N.Y. Masse, J.D. Simeral, J. Vogel, S. Haddadin, J. Liu, S.S. Cash, P. van der Smagt, and J.P. Donoghue. "Reach and grasp by people with tetraplegia using a neurally controlled robotic arm". In: *Nature* 485 (2012), pp. 372–377.
DOI: 10.1038/nature11076.

[90] G. Hoffman and C. Breazeal. "Collaboration in human-robot teams". In: *In Proc. of the AIAA Intelligent Systems Technical Conference*. 2004.
URL: http://alumni.media.mit.edu/~guy/publications/HoffmanAIAA04.pdf.

[91] P.F. Hokayem and M.W. Spong. "Bilateral Teleoperation: An Historical Survey". In: *Automatica* 42.12 (2006), pp. 2035–2057.
DOI: 10.1016/j.automatica.2006.06.027.

[92] W.S. Howard and V. Kumar. "On the Stability of Grasped Objects". In: *IEEE Trans. on Robotics and Automation* 12.6 (1996), pp. 904–917.
DOI: 10.1109/70.544773.

[93] H. Hu, J. Li, Z. Xie, B. Wang, H. Liu, and G. Hirzinger. "A robot arm/hand teleoperation system with telepresence and shared control". In: *Proc. of the IEEE/ASME Int. Conf. on Advanced Intelligent Mechatronics*. July 2005, pp. 1312–1317.
DOI: 10.1109/AIM.2005.1511192.

[94] K. Hübner and D. Kragic. "Selection of robot pre-grasps using box-based shape approximation". In: *Proc. IEEE/RSJ Int. Conf. on Intelligent Robots and Systems (IROS)*. Sept. 2008, pp. 1765–1770.
DOI: 10.1109/IROS.2008.4650722.

[95] iRobot. *710 Kobra*. Online; accessed 5-January-2015.
URL: http://www.irobot.com/For-Defense-and-Security/Robots/710-Kobra.aspx#PublicSafety.

[96] M. Jeannerod. "The Timing of Natural Prehension Movements". In: *Journal of Motor Behavior* 16.3 (1984), pp. 235–254.
DOI: 10.1080/00222895.1984.10735319.

[97] H. Jeong and J. Cheong. "In-hand Rolling Motion Planning using Independent Contact Region (ICR) with Guaranteed Grasp Quality Margin". In: *Proc. IEEE Int. Conf. on Robotics and Automation (ICRA)*. 2013, pp. 3239–3244.
DOI: 10.1109/ICRA.2013.6631028.

[98] M. Johnson, J.M. Bradshaw, R.R. Hoffman, P.J. Feltovich, and D.D. Woods. "Seven Cardinal Virtues of Human-Machine Teamwork: Examples from the DARPA Robotic Challenge". In: *IEEE Intelligent Systems* 29.6 (Nov. 2014), pp. 74–80.
DOI: 10.1109/MIS.2014.100.

[99] M.D. Johnston and K.J. Rabe. "Integrated planning for telepresence with time delays". In: *Proc. IEEE Int. Conf. on Space Mission Challenges for Information Technology*. 2006.
DOI: 10.1109/SMC-IT.2006.39.

[100] D. Kaber, E. Onal, and M. Endsley. "Design of automation for telerobots and the effect on performance, operator situation awareness and subjective workload". In: *Human Factors and Ergonomics in Manufacturing* 10 (2000), pp. 409–430.
DOI: 10.1002/1520-6564(200023)10:4<409::AID-HFM4>3.0.CO;2-V.

[101] C. van de Kamp and F.T. Zaal. "Prehension is really reaching and grasping". In: *Experimental Brain Research* 182.1 (2007), pp. 27–34.
DOI: 10.1007/s00221-007-0968-2.

[102] K. Khokar, R. Alqasemi, K.B. Reed, and R. Dubey. "Laser Assisted Combined Teleoperation and Autonomous Control". In: *Proc. ANS EPRRSD Robotics and remote Systems for Hazardous Environments, Emergency Preparedness & Response*. Aug. 2011.
URL: http://reedlab.eng.usf.edu/publications/Khokar2011laser.pdf.

[103] K. Khokar, K.B. Reed, R. Alqasemi, and R. Dubey. "Laser-assisted telerobotic control for enhancing manipulation capabilities of persons with disabilities". In: *Proc. IEEE/RSJ Int. Conf. on Intelligent Robots and Systems (IROS)*. Oct. 2010, pp. 5139–5144.
DOI: 10.1109/IROS.2010.5653184.

[104] H.K. Kim, S.J. Biggs, D.W. Schloerb, J.M. Carmena, M.A. Lebedev, M.A. Nicolelis, and M.A. Srinivasan. "Continuous shared control for stabilizing reaching and grasping with brain-machine interfaces". In: *IEEE Trans Biomed Eng.* 53.6 (June 2006), pp. 1164–1173.
DOI: 10.1109/TBME.2006.870235.

[105] J. Kim, K. Iwamoto, J.J. Kuffner, Y. Ota, and N.S. Pollard. "Physically Based Grasp Quality Evaluation Under Pose Uncertainty". In: *IEEE Trans. on Robotics* 29.6 (Dec. 2013), pp. 1424–1439.
DOI: 10.1109/TRO.2013.2273846.

[106] F. Kobayashi, H. Nakamoto, F. Kojima, T. Maeda, N. Imamura, and H. Shirasawa. "Teleoperation of Universal Robot Hand with Pinching Force Stabilization". In: *Simulation and Modeling Related to Computational Science and Robotics Technology* 37 (2012), pp. 172–184.
DOI: 10.3233/978-1-61499-092-5-172.

[107] R. Konietschke, Ulrich Hagn, Mathias Nickl, Stefan Jorg, Andreas Tobergte, Georg Passig, U. Seibold, Luc Le-Tien, B. Kubler, Martin Groger, Florian Frohlich, Christian Rink, Alin Albu-Schäffer, Markus Grebenstein, T. Ortmaier, and G. Hirzinger. "The DLR MiroSurge - A robotic system for surgery". In: *Proc. IEEE Int. Conf. on Robotics and Automation (ICRA).* May 2009, pp. 1589–1590.
DOI: 10.1109/ROBOT.2009.5152361.

[108] R. Konietschke, A. Tobergte, C. Preusche, P. Tripicchio, E. Ruffaldi, S. Webel, and U. Bockholt. "A Multimodal Training Platform for Minimally Invasive Robotic Surgery". In: *Proc. IEEE Int. Workshop on Robot and Human Interactive Communication (ROMAN).* Sept. 2010, pp. 449–454. ISBN: 978-1-4244-7989-4.
DOI: 10.1109/ROMAN.2010.5598608.

[109] D. Kortenkamp, R.P. Bonasso, D. Ryan, and D. Schreckenghost. "Traded control with autonomous robots as mixed initiative interaction". In: *AAAI Symposium on Mixed Initiative Interaction.* 1997.
URL: http://www.aaai.org/Papers/Symposia/Spring/1997/SS-97-04/SS97-04-016.pdf.

[110] D. Kortenkamp, D. Keirn-Schreckenghost, and R.P. Bonasso. "Adjustable control autonomy for manned space flight". In: *Proc. of the IEEE Aerospace Conference.* Vol. 7. 2000, pp. 629–640.
DOI: 10.1109/AERO.2000.879330.

[111] I. Kossyk, J. Dorr, Lars Raschendorfer, and K. Kondak. "Usability of a virtual reality system based on a wearable haptic interface". In: *Proc. IEEE/RSJ Int. Conf. on Intelligent Robots and Systems (IROS).* Sept. 2011, pp. 3474–3479.
DOI: 10.1109/IROS.2011.6094839.

[112] P. Kremer, T. Wimböck, J. Artigas, S. Schätzle, K. Jöhl, F. Schmidt, C. Preusche, and G. Hirzinger. "Multimodal telepresent control of DLR's Rollin' Justin". In: *Proc. IEEE Int. Conf. on Robotics and Automation (ICRA).* May 2009.
DOI: 10.1109/ROBOT.2009.5152670.

[113] T. Krueger and A. Schiele. "Exoskeleton Control of the Robonaut through RAPID and ROS". In: *12th Symp. on Advanced Space Technologies for Robotics and Automation (ASTRA).* 2012.
URL: http://robotics.estec.esa.int/ASTRA/Astra2013/Papers/Krueger_Schiele_2810861.pdf.

[114] R. Krug, D. Dimitrov, K. Charusta, and B. Iliev. "On the efficient computation of independent contact regions for force closure grasps". In: *Proc. IEEE/RSJ Int. Conf. on Intelligent Robots and Systems (IROS).* 2010, pp. 586–591.
DOI: 10.1109/IROS.2010.5654380.

[115] A.B. Kuang, S. Payandeh, Bin Zheng, F. Henigman, and C.L. MacKenzie. "Assembling virtual fixtures for guidance in training environments". In: *Proc. Int. Symp. on Haptic Interfaces for Virtual Environment and Teleoperator Systems.* Mar. 2004, pp. 367–374.
DOI: 10.1109/HAPTIC.2004.1287223.

[116] A. Leeper, K. Hsiao, M. Ciocarlie, L. Takayama, and D. Gossow. "Strategies for human-in-the-loop robotic grasping". In: *Proc. ACM/IEEE Int. Conf. on Human-Robot Interaction (HRI).* 2012, pp. 1–8.
DOI: 10.1145/2157689.2157691.

[117] S.J. Levine, S. Schaffert, and N. Checka. "Natural User Interface for Robot Task Assignment". In: *Proc. Robotics: Science and Systems; Workshop on Human-Robot Collaboration for Industrial Manufacturing*. 2014.
URL: http://hci.cs.wisc.edu/workshops/RSS2014/wp-content/uploads/2013/12/levine2014natural.pdf.

[118] H. Li, C.D. Wickens, N. Sarter, and A. Sebok. "Stages and Levels of Automation in Support of Space Teleoperations". In: *Human Factors: J. of Human Factors and Ergonomics Society* 56.6 (2014), pp. 1050–1061.
DOI: 10.1177/0018720814522830.

[119] M.V. Liarokapis, P.K. Artemiadis, and K.J. Kyriakopoulos. "Mapping human to robot motion with functional anthropomorphism for teleoperation and telemanipulation with robot arm hand systems". In: *Proc. IEEE/RSJ Int. Conf. on Intelligent Robots and Systems (IROS)*. Nov. 2013, pp. 2075–2075.
DOI: 10.1109/IROS.2013.6696638.

[120] S. Lichiardopol. *A survey on teleoperation*. 2007.
URL: http://www.mate.tue.nl/mate/pdfs/8832.pdf.

[121] N.Y. Lii, Z. Chen, B. Pleintinger, C. Borst, G. Hirzinger, and A. Schiele. "Toward understanding the effects of visual- and force- feedback on robotic hand grasping performance for space teleoperation". In: *Proc. IEEE/RSJ Int. Conf. on Intelligent Robots and Systems (IROS)*. Oct. 2010.
DOI: 10.1109/IROS.2010.5650186.

[122] N.Y. Lii, Z. Chen, M.A. Roa, A. Maier, B. Pleintinger, and C. Borst. "Toward a task space framework for gesture commanded telemanipulation." In: *Proc. IEEE Int. Workshop on Robot and Human Interactive Communication (ROMAN)*. 2012, pp. 925–932.
DOI: 10.1109/ROMAN.2012.6343869.

[123] V. Lippiello, B. Siciliano, and L. Villani. "A Grasping Force Optimization Algorithm for Multiarm Robots With Multifingered Hands". In: *IEEE Trans. on Robotics* 29.1 (Feb. 2013), pp. 55–67.
DOI: 10.1109/TRO.2012.2212633.

[124] H. Liu, K. Wu, P. Meusel, N. Seitz, G. Hirzinger, M.H. Jin, Y.W. Liu, S.W. Fan, T. Lan, and Z.P. Chen. "Multisensory Five-Finger Dexterous Hand: The DLR/HIT Hand II". In: *Proc. IEEE/RSJ Int. Conf. on Intelligent Robots and Systems (IROS)*. 2008, pp. 3692–3697.
DOI: 10.1109/IROS.2008.4650624.

[125] Y.-C. Liu and N. Chopra. "Semi-autonomous teleoperation in task space with redundant slave robot under communication delays". In: *Proc. IEEE/RSJ Int. Conf. on Intelligent Robots and Systems (IROS)*. Sept. 2011, pp. 679–684.
DOI: 10.1109/IROS.2011.6095114.

[126] Y.H. Liu. "Qualitative Test and Force Optimization of 3-D Frictional Form-Closure Grasps Using Linear Programming". In: *IEEE Trans. on Robotics and Automation* 15.1 (1999), pp. 163–173.
DOI: 10.1109/70.744611.

[127] Y.H. Liu. "Computing n-Finger Form-Closure Grasps on Polygonal Objects". In: *Int. J. Robotics Research* 19.2 (2000), pp. 149–158.
DOI: 10.1177/02783640022066798.

[128] Deutsches Zentrum für Luft- und Raumfahrt (DLR). *DLR at ILA 2010: Spacerobot SpaceJustin Meets Journalist.* accessed 10-Oct-2014.
URL: http://www.dlr.de/media/desktopdefault.aspx/tabid-4996//8427_read-15390.

[129] C.L. MacKenzie and T. Iberall. *The Grasping Hand.* North-Holland, 1994. ISBN: 9780444817464.
URL: http://store.elsevier.com/product.jsp?isbn=9780444817464.

[130] J. Mainprice, C. Phillips-Grafflin, H.B. Suay, N. Alunni, D. Lofaro, D. Berenson, S. Chernova, R.W. Lindeman, and P. Oh. "From autonomy to cooperative traded control of humanoid manipulation tasks with unreliable communication: System design and lessons learned". In: *Proc. IEEE/RSJ Int. Conf. on Intelligent Robots and Systems (IROS).* Sept. 2014, pp. 3767–3774.
DOI: 10.1109/IROS.2014.6943091.

[131] P. Malysz and S. Sirouspour. "Maneuverability and grasping experiments in teleoperation of nonholonomic/twin-armed robots". In: *Proc. IEEE Haptics Symposium.* Mar. 2012, pp. 203–209.
DOI: 10.1109/HAPTIC.2012.6183792.

[132] P. Marayong and A.M. Okamura. "Speed-Accuracy Characteristics of Human-Machine Cooperative Manipulation Using Virtual Fixtures With Variable Admittance". In: *Human Factors: J. of Human Factors and Ergonomics Society* 46.3 (2004), pp. 518–532.
DOI: 10.1518/hfes.46.3.518.50400.

[133] M. Markovic, S. Dosen, C. Cipriani, D. Popovic, and D. Farina. "Stereovision and augmented reality for closed-loop control of grasping in hand prostheses". In: *J. of Neural Engineering* 11.4 (2014).
DOI: 10.1088/1741-2560/11/4/046001.

[134] Z.-C. Márton, D. Pangercic, N. Blodow, and M. Beetz. "Combined 2D–3D Categorization and Classification for Multimodal Perception Systems". In: *Int. J. Robotics Research* 30.11 (Sept. 2011), pp. 1378–1402.
DOI: 10.1177/0278364911415897.

[135] M. Mast, M. Burmester, K. Krüger, S. Fatikow, G. Arbeiter, B. Graf, G. Kronreif, L. Pigini, D. Facal, and R. Qiu. "User-Centered Design of a Dynamic-Autonomy Remote Interaction Concept for Manipulation-Capable Robots to Assist Elderly People in the Home". In: *Journal of Human-Robot Interaction* 1.1 (2012).
URL: http://www.humanrobotinteraction.org/journal/index.php/HRI/article/view/19.

[136] P. Michelman and P. Allen. "Shared autonomy in a robot hand teleoperation system". In: *Proc. IEEE/RSJ Int. Conf. on Intelligent Robots and Systems (IROS).* Vol. 1. 1994, pp. 253–259.
DOI: 10.1109/IROS.1994.407383.

[137] Microsoft. *Xbox.* Online; accessed 5-January-2015.
URL: http://www.xbox.com/de-DE/Xbox360/Accessories/Controllers.

[138] P. Milgram, A. Rastogi, and J.J. Grodski. "Telerobotic control using augmented reality". In: *Proc. IEEE Int. Workshop on Robot and Human Interactive Communication (ROMAN).* July 1995, pp. 21–29.
DOI: 10.1109/ROMAN.1995.531930.

[139] A.T. Miller and P.K. Allen. "GraspIt! A Versatile Simulator for Robotic Grasping". In: *IEEE Robotics and Automation Magazine* 11.4 (2004), pp. 110–122.
DOI: 10.1109/MRA.2004.1371616.

[140] A.T. Miller, S. Knoop, H.I. Christensen, and P.K. Allen. "Automatic Grasp Planning Using Shape Primitives". In: *Proc. IEEE Int. Conf. on Robotics and Automation (ICRA)*. 2003, pp. 1824–1829.
DOI: 10.1109/ROBOT.2003.1241860.

[141] A. Morales, T. Asfour, P. Azad, S. Knoop, and R. Dillmann. "Integrated Grasp Planning and Visual Object Localization For a Humanoid Robot with Five-Fingered Hands". In: *Proc. IEEE/RSJ Int. Conf. on Intelligent Robots and Systems (IROS)*. Oct. 2006, pp. 5663–5668.
DOI: 10.1109/IROS.2006.282367.

[142] R. Murphy. *Introduction to AI Robotics*. A Bradford book. MIT Press, 2000. ISBN: 9780262133838.
URL: http://books.google.de/books?id=RVlnL_X6FrwC.

[143] R.M. Murray. "Recent research in cooperative control of multivehicle systems". In: *J. Dynamic Systems, Measurement, and Control* 129 (2007), pp. 571–583.
URL: http://users.cms.caltech.edu/~murray/preprints/mur07-dsmc.pdf.

[144] R.M. Murray, Z. Li, and S. Sastry. *A Mathematical Introduction to Robotic Manipulation*. Boca Raton, Florida: CRC Press, 1994.

[145] S. Muszynski, J. Stuckler, and S. Behnke. "Adjustable autonomy for mobile teleoperation of personal service robots". In: *Proc. IEEE Int. Workshop on Robot and Human Interactive Communication (ROMAN)*. Sept. 2012, pp. 933–940.
DOI: 10.1109/ROMAN.2012.6343870.

[146] NASA. *Virtual Reality Training*. Online; accessed 5-January-2015.
URL: http://www.nasa.gov/audience/forstudents/5-8/features/F_STS-118_Cosmic_Corridor_Text_prt.htm.

[147] V.D. Nguyen. "Constructing stable grasps in 3D". In: *Proc. IEEE Int. Conf. on Robotics and Automation (ICRA)*. Vol. 4. Mar. 1987, pp. 234–239.
DOI: 10.1109/ROBOT.1987.1088008.

[148] V.D. Nguyen. "Constructing Force-Closure Grasps". In: *Int. J. Robotics Research* 7.3 (1988), pp. 3–16.
DOI: 10.1177/027836498800700301.

[149] G. Niemeyer, C. Preusche, and G. Hirzinger. "Telerobotics". In: *Springer Handbook of Robotics*. Ed. by Bruno Siciliano and Oussama Khatib. Springer Berlin Heidelberg, 2008, pp. 741–757. ISBN: 978-3-540-23957-4.
DOI: 10.1007/978-3-540-30301-5_32.

[150] Novint. *Novint Falcon*. Online; accessed 5-January-2015.
URL: http://www.novint.com/index.php/novintfalcon.

[151] M. O'Malley, A. Gupta, M. Gen, and Y. Li. "Shared Control in Haptic Systems for Performance Enhancement and Training". In: *J. Dynamic Systems, Measurement, and Control* 128 (2006).
URL: http://me.rice.edu/~abhi/pubs/JDSMC06.pdf.

[152] M.K. O'Malley and R.O. Ambrose. "Haptic feedback applications for Robonaut". In: *Industrial Robot: An International Journal* 30.6 (2003), pp. 531–542.
DOI: 10.1108/01439910310506800.

[153] Robot Operating System (ROS) Open Source Robotic Foundation. *3D Robot Visualizer (RViz)*. Online; accessed 31-January-2015.
URL: http://wiki.ros.org/rviz.

[154] N. Padoy and G.D. Hager. "Human-Machine Collaborative surgery using learned models". In: *Proc. IEEE Int. Conf. on Robotics and Automation (ICRA)*. May 2011, pp. 5285–5292. DOI: 10.1109/ICRA.2011.5980250.

[155] D. Pallaschke and J. Rosenmüller. "Computing the Minkowski Sum of Prisms". In: *Journal of Global Optimization* 35.2 (2006), pp. 321–341. ISSN: 0925-5001. DOI: 10.1007/s10898-005-3841-5.

[156] R. Parasuraman, T.B. Sheridan, and Christopher D. Wickens. "A model for types and levels of human interaction with automation". In: *IEEE Trans. on Systems, Man, and Cybernetics; Part A: Systems and Humans* 30.3 (May 2000), pp. 286–297. DOI: 10.1109/3468.844354.

[157] L.E. Parker and J.V. Draper. "Robotics applications in maintenance and repair". In: *Handbook of Industrial Robotics* (1998), pp. 1023–1036. URL: https://books.google.de/books?id=7od4alFKfNMC.

[158] Parrot. *AR Drone 2*. Online; accessed 5-January-2015. 2014. URL: http://ardrone2.parrot.com/.

[159] A. Peer, S. Einenkel, and M. Buss. "Multi-fingered telemanipulation - mapping of a human hand to a three finger gripper". In: *Proc. IEEE Int. Workshop on Robot and Human Interactive Communication (ROMAN)*. Aug. 2008, pp. 465–470. DOI: 10.1109/ROMAN.2008.4600710.

[160] C. Perez Quintero, R.T. Fomena, A. Shademan, O. Ramirez, and M. Jagersand. "Interactive Teleoperation Interface for Semi-autonomous Control of Robot Arms". In: *Proc. Canadian Conf. on Computer and Robot Vision (CRV)*. May 2014, pp. 357–363. DOI: 10.1109/CRV.2014.55.

[161] N.P. Pernalete, W. Yu, R.V. Dubey, and W.A Moreno. "Augmentation of manipulation capabilities of persons with disabilities using scaled teleoperation". In: *Proc. IEEE/RSJ Int. Conf. on Intelligent Robots and Systems (IROS)*. Vol. 2. 2002, pp. 1517–1522. DOI: 10.1109/IRDS.2002.1043970.

[162] C. Phillips-Grafflin, N. Alunni, H. Suay, J. Mainprice, D. Lofaro, D. Berenson, S. Chernova, R.W. Lindeman, and P. Oh. "Toward a user-guided manipulation framework for high-DOF robots with limited communication". In: *Intelligent Service Robotics* 7.3 (2014), pp. 121–131. ISSN: 1861-2776. DOI: 10.1007/s11370-014-0156-8.

[163] R.M. Pierce and K.J. Kuchenbecker. "A data-driven method for determining natural human-robot motion mappings in teleoperation". In: *Proc. IEEE RAS EMBS Int. Conf. on Biomedical Robotics and Biomechatronics*. June 2012, pp. 169–176. DOI: 10.1109/BioRob.2012.6290927.

[164] N.S. Pollard. "Closure and Quality Equivalence for Efficient Synthesis of Grasps from Examples". In: *Int. J. Robotics Research* 23.6 (2004), pp. 595–614. DOI: 10.1177/0278364904044402.

[165] J. Ponce, S. Sullivan, A. Sudsang, J.D. Boissonat, and J.P. Merlet. "On Computing Four-Finger Equilibrium and Force-Closure Grasps of Polyhedral Objects". In: *Int. J. Robotics Research* 16.1 (1997), pp. 11–35. DOI: 10.1177/027836499701600102.

[166] O. Porges, T. Stouraitis, C. Borst, and M.A. Roa. "Reachability and Capability Analysis for Manipulation Tasks". In: *ROBOT2013: First Iberian Robotics Conference*. Ed. by M.A. Armada, A. Sanfeliu, and M. Ferre. Vol. 253. Advances in Intelligent Systems and Computing. Springer International Publishing, 2014, pp. 703–718. ISBN: 978-3-319-03652-6. DOI: 10.1007/978-3-319-03653-3_50.

[167] R. Prada and S. Payandeh. "A study on design and analysis of virtual fixtures for cutting in training environments". In: *IEEE World Haptics Conf., Joint Eurohaptics Conf. and Symp. on Haptic Interfaces for Virtual Environment and Teleoperator Systems*. Mar. 2005, pp. 375–380. DOI: 10.1109/WHC.2005.19.

[168] D. Prattichizzo and J.C. Trinkle. "Chapter 28: Grasping". In: *Springer Handbook of Robotics*. Ed. by Bruno Siciliano and Oussama Khatib. Springer Berlin Heidelberg, 2008, pp. 671–700. ISBN: 978-3-540-23957-4.

[169] F. Rehnmark, W. Bluethmann, J. Mehling, R.O. Ambrose, M. Diftler, M. Chu, and R. Necessary. "Robonaut: the 'short list' of technology hurdles". In: *Computer* 38.1 (Jan. 2005), pp. 28–37. DOI: 10.1109/MC.2005.32.

[170] M. A. Roa and R. Suárez. "Influence of contact types and uncertainties in the computation of Independent Contact Regions". In: *Proc. IEEE Int. Conf. on Robotics and Automation (ICRA)*. 2011, pp. 3317–3323. DOI: 10.1109/ICRA.2011.5980334.

[171] M.A. Roa, M.J. Argus, D. Leidner, C. Borst, and G. Hirzinger. "Power grasp planning for anthropomorphic robot hands". In: *Proc. IEEE Int. Conf. on Robotics and Automation (ICRA)*. May 2012, pp. 563–569. DOI: 10.1109/ICRA.2012.6225068.

[172] M.A. Roa and R. Suarez. "Computation of Independent Contact Regions for Grasping 3-D Objects". In: *IEEE Trans. on Robotics* 25.4 (2009), pp. 839–850. DOI: 10.1109/TRO.2009.2020351.

[173] M.A. Roa and R. Suarez. "Regrasp Planning in the Grasp Space Using Independent Regions". In: *Proc. IEEE/RSJ Int. Conf. on Intelligent Robots and Systems (IROS)*. 2009, pp. 1823–1829. DOI: 10.1109/IROS.2009.5354479.

[174] M.A. Roa and R. Suárez. "Grasp quality measures: review and performance". In: *Autonomous Robots* 38.1 (2015), pp. 65–88. DOI: 10.1007/s10514-014-9402-3.

[175] Double Robotics. *Telepresence robot*. Online; accessed 5-January-2015. URL: http://www.doublerobotics.com/.

[176] Schilling Robotics. *UHD-III ROV*. Online; accessed 5-January-2015. URL: http://www.fmctechnologies.com/en/SchillingRobotics/Technologies/Schilling-UHD-ROV.aspx.

[177] R.N. Rohling and J.M. Hollerbach. "Optimized fingertip mapping for teleoperation of dextrous robot hands". In: *Proc. IEEE Int. Conf. on Robotics and Automation (ICRA)*. Vol. 3. May 1993, pp. 769–775. DOI: 10.1109/ROBOT.1993.292238.

[178] C. Rosales, R. Suarez, M. Gabiccini, and A. Bicchi. "On the synthesis of feasible and prehensile robotic grasps". In: *Proc. IEEE Int. Conf. on Robotics and Automation (ICRA)*. May 2012, pp. 550–556.
DOI: 10.1109/ICRA.2012.6225238.

[179] J. Rosell, X. Sierra, L. Palomo, and R. Suarez. "Finding grasping configurations of a dexterous hand and an industrial robot". In: *Proc. IEEE Int. Conf. on Robotics and Automation (ICRA)*. 2005, pp. 1178–1183.
DOI: 10.1109/ROBOT.2005.1570275.

[180] J. Rosell, R. Suárez, and A. Pérez. "Safe Teleoperation of a Dual Hand-Arm Robotic System." In: *ROBOT2013: First Iberian Robotics Conference*. 2013, pp. 615–630.
DOI: 10.1007/978-3-319-03653-3_44.

[181] M. Sagardia and T. Hulin. "Fast and accurate distance, penetration, and collision queries using point-sphere trees and distance fields". In: *SIGGRAPH Posters*. 2013, p. 83.
URL: http://elib.dlr.de/82751/.

[182] M. Sagardia, T. Hulin, C. Preusche, and G. Hirzinger. "Improvements of the Voxmap-Pointshell Algorithm - Fast Generation of Haptic Data Structures". In: *Proc. 53rd Int. Wissenschaftliches Kolloquium*. 2008.

[183] G. Salvietti, M. Malvezzi, G. Gioioso, and D. Prattichizzo. "On the use of homogeneous transformations to map human hand movements onto robotic hands". In: *Proc. IEEE Int. Conf. on Robotics and Automation (ICRA)*. May 2014, pp. 5352–5357.
DOI: 10.1109/ICRA.2014.6907646.

[184] A. Saxena, J. Driemeyer, and A.Y. Ng. "Robotic Grasping of Novel Objects using Vision". In: *The International Journal of Robotics Research* 27.2 (2008), pp. 157–173.
DOI: 10.1177/0278364907087172.

[185] S. Schätzle, T. Ende, T. Wüsthoff, and C. Preusche. "VibroTac: An ergonomic and versatile usable vibrotactile feedback device". In: *Proc. IEEE Int. Workshop on Robot and Human Interactive Communication (ROMAN)*. Sept. 2010, pp. 670–675.
DOI: 10.1109/ROMAN.2010.5598694.

[186] R. Scheuchenpflug. "Measuring presence in virtual environments". In: *HCI International*. 2001, pp. 56–58.

[187] T. Schlegl, M. Buss, T. Omata, and G.K. Schmidt. "Fast dextrous re-grasping with optimal contact forces and contact sensor-based impedance control". In: *Proc. IEEE Int. Conf. on Robotics and Automation (ICRA)*. Vol. 1. 2001, pp. 103–108.
DOI: 10.1109/ROBOT.2001.932537.

[188] V. Schmirgel, U.E. Zimmermann, T. Hulin, and C. Preusche. "Position Paper: Human Skills for Programming-by-Demonstration of Robots". In: *beyond movement*. Ed. by Alinea editrice s.r.l. 2008, pp. 144–167.
URL: http://elib.dlr.de/55618/.

[189] Sensable. *Phantom Omni*. Online; accessed 5-January-2015.
URL: http://www.dentsable.com/haptic-phantom-omni.htm.

[190] A. Settimi, C. Pavan, V. Varricchio, M. Ferrati, E.M. Hoffman, A. Rocchi, K. Melo, N.G. Tsagarakis, and A. Bicchi. "A modular approach for remote operation of humanoid robots in search and rescue scenarios". In: *Modelling and Simulation for Autonomous Systems: First Int. Workshop, (MESAS)* (2014).
DOI: 10.1007/978-3-319-13823-7_18.

[191] T.B. Sheridan. *Telerobotics, Automation, and Human Supervisory Control*. MIT Press, 1992. ISBN: 9780262193160.
URL: http://books.google.de/books?id=eu41_M2Do9oC.

[192] T.B. Sheridan. "Adaptive Automation, Level of Automation, Allocation Authority, Supervisory Control, and Adaptive Control: Distinctions and Modes of Adaptation". In: *IEEE Trans. on Systems, Man, and Cybernetics; Part A: Systems and Humans* 41.4 (July 2011), pp. 662–667.
DOI: 10.1109/TSMCA.2010.2093888.

[193] K.B. Shimoga. "A survey of perceptual feedback issues in dexterous telemanipulation. I. Finger force feedback". In: *Proc. on the Virtual Reality Annual International Symposium*. 1993, pp. 263–270.

[194] K. Shiratsuchi, K. Kawata, E.V. Poorten, and Y. Yokokohji. "Design and Evaluation of a Telepresence Vision System for Manipulation Tasks". In: *Proc. IEEE Int. Conf. on Robotics and Automation (ICRA)*. Apr. 2007, pp. 4313–4318.
DOI: 10.1109/ROBOT.2007.364143.

[195] K. Shoemake. "Animating Rotation with Quaternion Curves". In: *SIGGRAPH Comput. Graph.* 19.3 (July 1985), pp. 245–254.
DOI: 10.1145/325165.325242.

[196] J.B.J. Smeets and E. Brenner. "A new view on grasping". In: *Motor Control* 3.3 (1999).
URL: http://personal.fbw.vu.nl/jsmeets/1999/NewViewGrasping-MC99.pdf.

[197] G. Srimathveeravalli, V. Gourishankar, and T. Kesavadas. "Comparative Study: Virtual Fixtures and Shared Control for Rehabilitation of Fine Motor Skills". In: *IEEE World Haptics Conf., Joint Eurohaptics Conf. and Symp. on Haptic Interfaces for Virtual Environment and Teleoperator Systems*. Mar. 2007, pp. 304–309.
DOI: 10.1109/WHC.2007.31.

[198] N. Stefanov, C. Passenberg, A. Peer, and M. Buss. "Design and Evaluation of a Haptic Computer-Assistant for Telemanipulation Tasks". In: *IEEE Trans. on Human-Machine Systems* 43.4 (July 2013), pp. 385–397.
DOI: 10.1109/TSMC.2013.2257743.

[199] M.R. Stein and R.P. Paul. "Operator interaction, for time-delayed teleoperation, with a behavior-based controller". In: *Proc. IEEE Int. Conf. on Robotics and Automation (ICRA)*. Vol. 1. May 1994, pp. 231–236.
DOI: 10.1109/ROBOT.1994.350983.

[200] M.R. Stein, R.P. Paul, P.S. Schenker, and E.D. Paljug. "A cross-country teleprogramming experiment". In: *Proc. IEEE/RSJ Int. Conf. Intelligent Robots and Systems. Human Robot Interaction and Cooperative Robots*. Vol. 2. Aug. 1995, pp. 21–26.
DOI: 10.1109/IROS.1995.526133.

[201] J. Steuer. "Defining Virtual Reality: Dimensions Determining Telepresence". In: *Journal of Communication* 42.4 (1992), pp. 73–93.
DOI: 10.1111/j.1460-2466.1992.tb00812.x.

[202] M. Strandberg. "A grasp evaluation procedure based on disturbance forces". In: *Proc. IEEE/RSJ Int. Conf. on Intelligent Robots and Systems (IROS)*. Vol. 2. 2002, pp. 1699–1704.
DOI: 10.1109/IRDS.2002.1044000.

[203] R. Suarez, M.A. Roa, and J. Cornell. *Grasp Quality Measures*. Tech. rep. IOC-DT-P-2006-10. Technical University of Catalunya, 2006.
URL: https://personalrobotics.ri.cmu.edu/files/courses/papers/SuarezEtal06.pdf.

[204] Intuitive Surgical. *DaVinci Surgical System*. Online; accessed 5-January-2015.
URL: http://intuitivesurgical.com/company/media/images/davinci_si_images.html.

[205] CyberGlove Systems. *CyberForce*. Online; accessed 5-January-2015.
URL: http://www.cyberglovesystems.com/products/cyberforce/overview.

[206] Y.P. Toh, S. Huang, J. Lin, M. Bajzek, G. Zeglin, and N.S. Pollard. "Dexterous telemanipulation with a multi-touch interface". In: *Proc. IEEE-RAS Int. Conf. on Humanoid Robots*. Nov. 2012, pp. 270–277.
DOI: 10.1109/HUMANOIDS.2012.6651531.

[207] W.T. Townsend. "The Barrett Hand Grasper - Programmably Flexible Part Handling and Assembly". In: *Industrial Robot* 27.3 (2000), pp. 181–188.

[208] K.M. Tsui, D.-J. Kim, A. Behal, D. Kontak, and H.A. Yanco. ""I Want That": Human-in-the-Loop Control of a Wheelchair-Mounted Robotic Arm". In: *Applied Bionics and Biomechanics* 8.1 (2011), pp. 127–147.
URL: http://dx.doi.org/10.3233/ABB-2011-0004.

[209] M.L. Turner, R.P. Findley, W.B Griffin, M.R. Cutkosky, and D.H. Gomez. "Development and Testing of a Telemanipulation System with Arm and Hand Motion". In: *Proc. ASME Dynamic Systems and Control Division (Symposium on Haptic Interfaces for Virtual Environments and Teleoperators)*. 2000, pp. 1057–1063.

[210] Defense Advanced Research Projects Agency (DARPA) United States Department of Defense. *The DARPA robotics challenge (DRC)*. Online; accessed 26-January-2015.
URL: http://www.theroboticschallenge.org/.

[211] N. Vahrenkamp, M. Kröhnert, S. Ulbrich, T. Asfour, G. Metta, R. Dillmann, and G. Sandini. "Simox: A Robotics Toolbox for Simulation, Motion and Grasp Planning". In: *Intelligent Autonomous Systems 12*. Ed. by S. Lee, H. Cho, K.-J. Yoon, and J. Lee. Vol. 193. Advances in Intelligent Systems and Computing. Springer Berlin Heidelberg, 2013, pp. 585–594. ISBN: 978-3-642-33925-7.
DOI: 10.1007/978-3-642-33926-4_55.

[212] H.S. Vitense. "Multimodal Feedback: An Assessment of Performance and Mental Workload". In: *Ergonomics* 46 (2003), pp. 68–87.
DOI: 10.1080/00140130303534.

[213] J. Vogel, J. Bayer, and P. van der Smagt. "Continuous robot control using surface electromyography of atrophic muscles". In: *Proc. IEEE/RSJ Int. Conf. on Intelligent Robots and Systems (IROS)*. IEEE. 2013, pp. 845–850.
DOI: 10.1109/IROS.2013.6696449.

[214] J. Vogel, S. Haddadin, B. Jarosiewicz, J.D. Simeral, D. Bacher, L.R. Hochberg, J.P. Donoghue, and P. van der Smagt. "An assistive decision-and-control architecture for force-sensitive hand-arm systems driven by human-machine interfaces". In: *Int. J. Robotics Research* (2015).
DOI: 10.1177/0278364914561535.

[215] R. Volcic and F. Domini. "The visibility of contact points influences grasping movements". In: *Experimental Brain Research* 232.9 (2014), pp. 2997–3005.
DOI: 10.1007/s00221-014-3978-x.

[216] B. Weber, M. Sagardia, T. Hulin, and C. Preusche. "Vibrotactile, Visual and Force Feedback of Collisions in Virtual Environments: Effects on Performance, Mental Workload and Spatial Orientation". In: *R. Shumaker (Ed.): Virtual, Augmented and Mixed Reality /HCII 2013, Part I, LNCS 8021*. 2013, pp. 241–250.
DOI: 10.1007/978-3-642-39405-8_28.

[217] J. Weisz, C. Elvezio, and P.K. Allen. "A user interface for assistive grasping". In: *Proc. IEEE/RSJ Int. Conf. on Intelligent Robots and Systems (IROS)*. Nov. 2013, pp. 3216–3221.
DOI: 10.1109/IROS.2013.6696813.

[218] Wikipedia. *Mouse (computing) — Wikipedia, The Free Encyclopedia*. Online; accessed 5-January-2015.
URL: http://en.wikipedia.org/w/index.php?title=Mouse_(computing)&oldid=639617952.

[219] Bob G. Witmer and Michael J. Singer. "Measuring Presence in Virtual Environments: A Presence Questionnaire". In: *Presence: Teleoper. and Virtual Environ.* 7.3 (June 1998), pp. 225–240.
URL: http://webstaff.itn.liu.se/~matco/TNM053/Papers/presence.pdf.

[220] T. Witzig, R. Jäkel, D. Pangercic, S. Osentoski, M. Zöllner, and R. Dillman. *Context Aware Shared Autonomy for Robotic Manipulation Tasks*. ICRA Workshop on Autonomous Learning. 2013.
URL: http://autonomous-learning.org/wp-content/uploads/13-ALW/paper_24.pdf.

[221] XSens. *MVN Biomeach Awinda*. Online; accessed 5-January-2015.
URL: https://www.xsens.com/products/mvn-biomech-awinda/.

[222] Z. Xue, J.M. Zoellner, and R. Dillmann. "Grasp Planning: Find the Contact Points". In: *Proc. IEEE Int. Conf. Robotics and Biomimetics*. 2007, pp. 835–840.
DOI: 10.1109/ROBIO.2007.4522271.

[223] Y. Yoshimura and R. Ozawa. "A supervisory control system for a multi-fingered robotic hand using datagloves and a haptic device". In: *Proc. IEEE/RSJ Int. Conf. on Intelligent Robots and Systems (IROS)*. Oct. 2012, pp. 5414–5419.
DOI: 10.1109/IROS.2012.6385486.

[224] E. You and K. Hauser. "Assisted Teleoperation Strategies for Aggressively Controlling a Robot Arm with 2D Input". In: *Proc. Robotics: Science and Systems*. July 2011.
URL: http://www.roboticsproceedings.org/rss07/p45.html.

[225] W. Yu, B.E. Fritz, N. Pernalete, M. Jurczyk, and R.V. Dubey. "Sensors assisted telemanipulation for maximizing manipulation capabilities of persons with disabilities". In: *Proc. Symp. on Haptic Interfaces for Virtual Environment and Teleoperator Systems*. Mar. 2003, pp. 295–301.
DOI: 10.1109/HAPTIC.2003.1191298.

[226] F. Zacharias, C. Borst, and G. Hirzinger. "Capturing Robot Workspace Structure: Representing Robot Capabilities". In: *Proc. IEEE/RSJ Int. Conf. on Intelligent Robots and Systems (IROS)*. Oct. 2007, pp. 3229–3236.
DOI: 10.1109/IROS.2007.4399105.

[227] F. Zacharias, C. Borst, and G. Hirzinger. "Object-Specific Grasp Maps for Use in Planning Manipulation Actions". In: *Proc. German Workshop on Robotics - WSR*. 2009, pp. 203–213.

[228] F. Zacharias, C. Borst, and G. Hirzinger. "Online Generation of Reachable Grasps for Dexterous Manipulation Using a Representation of the Reachable Workspace". In: *Proc. IEEE Int. Conf. Advanced Robotics*. 2009, pp. 3229–3236.

[229] F. Zacharias, C. Schlette, F. Schmidt, C. Borst, J. Rossmann, and G. Hirzinger. "Making planned paths look more human-like in humanoid robot manipulation planning". In: *Proc. IEEE Int. Conf. on Robotics and Automation (ICRA)*. 2011, pp. 1192–1198.
DOI: 10.1109/ICRA.2011.5979553.

[230] Y. Zheng and C.-M. Chew. "Distance Between a Point and a Convex Cone in n-Dimensional Space: Computation and Applications". In: *IEEE Trans. on Robotics* 25.6 (2009), pp. 1397–1412.
DOI: 10.1109/TRO.2009.2033333.

[231] Y. Zheng, M.C. Lin, and D. Manocha. "A fast n-dimensional ray-shooting algorithm for grasping force optimization". In: *Proc. IEEE Int. Conf. on Robotics and Automation (ICRA)*. 2010, pp. 1300–1305.
DOI: 10.1109/ROBOT.2010.5509297.

[232] Y. Zheng and W.H. Qian. "Linearizing the soft finger contact constraint with application to dynamic force distribution in multifingered grasping". In: *Science in China E: Engineering & Materials Science* 48.2 (2005), pp. 121–130.
DOI: 10.1360/04ye0210.

[233] Y. Zheng and K. Yamane. "Ray-Shooting Algorithms for Robotics". In: *Algorithmic Foundations of Robotics X*. Ed. by E. Frazzoli, T. Lozano-Perez, N. Roy, and D. Rus. Vol. 86. Springer Tracts in Advanced Robotics. Springer Berlin Heidelberg, 2013, pp. 105–121. ISBN: 978-3-642-36278-1.
DOI: 10.1007/978-3-642-36279-8_7.

[234] Yu Zheng and W.-H. Qian. "Improving grasp quality evaluation". In: *Robotics and Autonomous Systems* 57.6-7 (2009), pp. 665–673.
DOI: 10.1016/j.robot.2008.12.002.

[235] X. Zhu and H. Ding. "Planning Force-Closure Grasps on 3-D Objects". In: *Proc. IEEE Int. Conf. on Robotics and Automation (ICRA)*. 2004, pp. 1258–1263.
DOI: 10.1109/ROBOT.2004.1307997.

Own References

[236] C. Borst, F. Zacharias, F. Schmidt, D. Leidner, M.A. Roa, K. Hertkorn, G. Grunwald, P. Falco, C. Natale, and E. Maggio. "Observation and Execution". In: *Advanced Bimanual Manipulation*. Ed. by B. Siciliano. Vol. 80. Springer Tracts in Advanced Robotics. Springer Berlin Heidelberg, 2012, pp. 59–122. ISBN: 978-3-642-29040-4.
DOI: 10.1007/978-3-642-29041-1_2.

[237] S. Büttner, Z.-C. Márton, and K. Hertkorn. "Automatic Scene Parsing for Generic Object Descriptions using Shape Primitives". In: *Robotics and Autonomous Systems* (2015). to appear.

[238] C. Castellini, K. Hertkorn, M. Sagardia, D.S. Gonzalez, and M. Nowak. "A virtual piano-playing environment for rehabilitation based upon ultrasound imaging". In: *Proc. IEEE RAS EMBS Int. Conf. Biomedical Robotics and Biomechatronics*. Aug. 2014, pp. 548–554.
DOI: 10.1109/BIOROB.2014.6913835.

[239] R. Dörner, G. Matthys, M. Bogen, S. Rilling, A. Gerndt, J. Dodiya, K. Hertkorn, T. Hulin, J. Hummel, M. Sagardia, R. Wolff, T. Kühnert, G. Brunnett, H. Buchholz, L. Blum, C. Menk, C. Bade, W. Schreiber, M. Greiner, T. Alexander, M. Kleiber, G. Bruder, and F. Steinicke. "Fallbeispiele für VR/AR". In: *Virtual und Augmented Reality (VR / AR)*. Ed. by R. Dörner, W. Broll, P. Grimm, and B Jung. eXamen.press. Springer Berlin Heidelberg, 2013, pp. 295–326. ISBN: 978-3-642-28902-6.
DOI: 10.1007/978-3-642-28903-3_9.

[240] M. Gridseth, K. Hertkorn, and M. Jagersand. "On Visual Servoing to Improve Performance of Robotic Grasping". In: *Conf. on Computer and Robot Vision (CRV)*. June 2015, pp. 245–252.
DOI: 10.1109/CRV.2015.39.

[241] K. Hertkorn, T. Hulin, P. Kremer, C. Preusche, and G. Hirzinger. "Time Domain Passivity Control for multi-degree of freedom haptic devices with time delay". In: *Proc. IEEE Int. Conf. on Robotics and Automation (ICRA)*. May 2010, pp. 1313–1319.
DOI: 10.1109/ROBOT.2010.5509148.

[242] K. Hertkorn, M.A. Roa, and C. Borst. "Planning in-hand object manipulation with multifingered hands considering task constraints". In: *Proc. IEEE Int. Conf. on Robotics and Automation (ICRA)*. May 2013, pp. 617–624.
DOI: 10.1109/ICRA.2013.6630637.

[243] K. Hertkorn, M.A. Roa, M. Brucker, P. Kremer, and C. Borst. "Virtual Reality Support for Teleoperation Using Online Grasp Planning". In: *Proc. IEEE/RSJ Int. Conf. on Intelligent Robots and Systems (IROS)*. Nov. 2013, pp. 2074–2074.
DOI: 10.1109/IROS.2013.6696637.

[244] K. Hertkorn, M.A. Roa, C. Preusche, C. Borst, and G. Hirzinger. "Identification of contact formations: Resolving ambiguous force torque information". In: *Proc. IEEE Int. Conf. on Robotics and Automation (ICRA)*. May 2012, pp. 3278–3284.
DOI: 10.1109/ICRA.2012.6225148.

[245] K. Hertkorn, M.A. Roa, T. Wimböck, and C. Borst. "Computation of Realistic Contact Forces in Grasping". In: *Int. Conf. on Robotics Science and Systems (RSS): Workshop on Human vs. Robot Grasping*. July 2014.
URL: http://mobilemanipulation.org/rss2014/images/Abstracts/hertkorn14rssws.pdf.

[246] K. Hertkorn, M.A. Roa, T. Wimböck, and C. Borst. "Simultaneous and Realistic Contact and Force Planning in Grasping". In: *Proc. IEEE/RSJ Int. Conf. on Intelligent Robots and Systems (IROS)*. to appear. Oct. 2015.

[247] K. Hertkorn, B. Weber, P. Kremer, M.A. Roa, and C. Borst. "Assistance for Telepresence Using Online Grasp Planning". In: *Proc. IEEE-RAS Int. Conf. on Humanoid Robots*. Oct. 2013, pp. 507–513.
DOI: 10.1109/HUMANOIDS.2013.7030021.

[248] K. Hertkorn, B. Weber, P. Kremer, M.A. Roa, and C. Borst. "Evaluation of Assisted Grasping in Telepresence Systems". In: *IEEE Trans. on Haptics* (2015). submitted.

[249] T. Hulin, K. Hertkorn, P. Kremer, S. Schätzle, J. Artigas, M. Sagardia, F. Zacharias, and C. Preusche. "The DLR bimanual haptic device with optimized workspace". In: *Proc. IEEE Int. Conf. on Robotics and Automation (ICRA)*. May 2011, pp. 3441–3442.
DOI: 10.1109/ICRA.2011.5980066.

[250] T. Hulin, K. Hertkorn, and C. Preusche. "Interactive Features for Robot Viewers". In: *Intelligent Robotics and Applications*. Ed. by C.-Y. Su, S. Rakheja, and H. Liu. Vol. 7508. Lecture Notes in Computer Science. Springer Berlin Heidelberg, 2012, pp. 181–193. ISBN: 978-3-642-33502-0.
DOI: 10.1007/978-3-642-33503-7_19.

[251] M.A. Roa, K. Hertkorn, C. Borst, and G. Hirzinger. "Reachable Independent Contact Regions for precision grasps". In: *Proc. IEEE Int. Conf. on Robotics and Automation (ICRA)*. May 2011, pp. 5337–5343.
DOI: 10.1109/ICRA.2011.5980341.

[252] M.A. Roa, K. Hertkorn, B.-A. Dang-Vu, M. Sagardia, and C. Borst. "On the Computation of Reachable and Optimal Independent Contact Regions". In: *IEEE Trans. on Robotics* (2015). submitted.

[253] M.A. Roa, K. Hertkorn, F. Zacharias, C. Borst, and G. Hirzinger. "Graspability map: A tool for evaluating grasp capabilities". In: *Proc. IEEE/RSJ Int. Conf. on Intelligent Robots and Systems (IROS)*. Sept. 2011, pp. 1768–1774.
DOI: 10.1109/IROS.2011.6094957.

[254] M. Sagardia, K. Hertkorn, D.S. Gonzalez, and C. Castellini. "Ultrapiano: A novel human-machine interface applied to virtual reality". In: *Proc. IEEE Int. Conf. on Robotics and Automation (ICRA)*. May 2014, pp. 2089–2089.
DOI: 10.1109/ICRA.2014.6907142.

[255] M. Sagardia, K. Hertkorn, T. Hulin, R. Wolff, J. Hummell, J. Dodiya, and A. Gerndt. "An interactive virtual reality system for on-orbit servicing". In: *Proc. IEEE Virtual Reality (VR)*. Mar. 2013.
DOI: 10.1109/VR.2013.6549444.

[256] U. Scarcia, K. Hertkorn, C. Melchiorri, G. Palli, and T. Wimböck. "Local Online Planning of Coordinated Manipulation Motion". In: *Proc. IEEE Int. Conf. on Robotics and Automation (ICRA)*. May 2015, pp. 6081–6087.
DOI: 10.1109/ICRA.2015.7140052.

[257] T. Schmidt, K. Hertkorn, R. Newcombe, Z.-C. Márton, M. Suppa, and D. Fox. "Robust Real-time Tracking with Visual and Physical Constraints for Robot Manipulation". In: *Proc. IEEE Int. Conf. on Robotics and Automation (ICRA)*. May 2015, pp. 119–126.
DOI: 10.1109/ICRA.2015.7138989.

[258] J. Vogel, K. Hertkorn, R.U. Menon, and M.A. Roa. "Flexible, Semi-Autonomous Grasping for Assistive Robotics". In: *Proc. IEEE Int. Conf. on Robotics and Automation (ICRA)*. to appear. 2016.

[259] B. Weber, K. Hertkorn, N. Döllinger, and P. Kremer. "Visuelle Assistenz bei telepräsenten Objektmanipulationen mit einem humanoiden Roboter". In: *Tagungsband 55. Fachausschusssitzung Anthropotechnik*. 2013.